Mathematical Methods for Engineers and Geoscientists

T0237859

Olga Wälder

Mathematical Methods
for Engineers and Geoscientists

 Springer

Dr. Olga Wälder
TU Dresden
Inst. Kartographie
Mommsenstr. 13
01062 Dresden
Germany
Olga.Waelder@tu-dresden.de

ISBN: 978-3-642-09456-9   e-ISBN: 978-3-540-75301-8

Cover design: deblik, Berlin

Printed on acid-free paper

9 8 7 6 5 4 3 2 1

springer.com

To
My high-school teachers in mathematics,
and to the best results I ever obtained:
To my sons
Julian Michail,
Maximilian Albert, and
Jakob Benedikt

Contents

Chapter 1
Introduction

We start with a fun puzzle in mathematics and mathematical methods. How many corners does a four-dimensional cube have? Does such a thing exist, you ask? You may be a geoscientist or a philosopher. If your answer is: there are surely more than the eight corners there are for a three-dimensional cube, you are an engineer. If you know without hesitation that there are exactly sixteen corners and you can prove why, you are a mathematician.

To explain the goal of this book, I refer to Hersh (1997):

> The United States suffers from "innumeracy" in its general population, "math avoidance" among high-school students, and 50 percent failure among college calculus students. Causes include starvation budgets in the school, mental attrition by television, parents who don't like math. There's another, unrecognized cause of failure: misconception of the nature of mathematics.

I think the specific reference to the United States may be omitted. It is really a worldwide problem. Moreover, there is one more consequence of "math avoidance" and "misconception": good mathematical approaches are sometimes applied incorrectly. Particularly, the methods of statistics are often misused for different goals. Applying mathematical methods is similar to using nuclear power: the final results depend on the competence of the user. I try to convince my readers to apply the "energy" of mathematics with consideration.

This book does not require special knowledge of pure mathematics. Equations and calculations are mostly rooted in high-school algebra, and the reader needs only a healthy human mind in order to understand them. This book is unusual in one sense, as unlike most mathematics books, it starts with some real problems and presentations and discussions follow.

This volume does not provide an invincible solution for the problem of "misconception" nor does it offer suggestions for classroom practice. It can assist in the education of engineers and geoscientists by helping them to understand the usefulness of diverse mathematical approaches. My first priority is to present these approaches in language that an engineer or a geoscientist can understand. I try to explain the mathematical methods as I do in my lectures for nonmathematicians, something akin to looking through special spectacles at so-called *mathematical reality*. In 1992 the great British mathematician G. H. Hardy in *A Mathematician's Apology* wrote about this in an amusing way:

> A chair or a star is not in the least like what it seems to be; the more we think of it, the fuzzier its outlines become in the haze of sensation which surrounds it; but "2" or "317" has nothing to do with sensation, and its properties stand out the more clearly the more closely we scrutinize it. It may be that modern physics fits best into some framework of idealistic philosophy. I do not believe it, but there are eminent physicists who say so. Pure mathematics, on the other hand, seems to me a rock on which all idealism founders: 317 is prime, not because we think so, or because our minds are shaped in one way rather than another, but because it is so, because mathematical reality is built that way.

We avoid the temptation to try and convince readers that mathematical reality does exist, as any discussion on this issue is similar to discussions about God. One either believes or does not believe. This is a special space, a cosmos where pure mathematicians live and work. However, some methods and approaches that are developed in this cosmos are undoubtedly useful for engineers and geoscientists.

This book is structured simply. Chapter 1 offers an introduction that is seldom read and nearly never finished. Chapter 2 presents a rough overview of the basic principles of mathematical modeling. This topic is not like a biscuit recipe. Really, there are infinitely many different tasks and correspondingly different kinds of modeling, so we discuss some common rules for mathematical modeling. Chapter 3, the primary focus of the book, is substantial and deals with some real problems. Here, in contrast to pure mathematics, an unusual method for engineers and geoscientists— an acceptable kind of presentation of mathematical approaches is used: Proceeding from practical problems we obtain the corresponding solutions by applying different mathematical methods. Discussing all real problems is impossible and would exceed the limits of a book. We present some practical problems, the kind that usually cause engineers and geoscientists to start to plow nervously through a pile of mathematics texts.

In Chap. 3, the basic idea of each mathematical approach is explained in a comprehensible way. First, a simple example is calculated. Second, a common rule for applying the approach is given and framed by a box. Note that a reader is also gently prepared for accessing the special "hard" mathematical literature on the corresponding topic. Advantages and disadvantages of certain approaches and the relationships among them are discussed. Important equations are numbered by chapter. For example (3-5) means equation 5 from Chap. 3. Internal equations used within examples are numbered starting with * and renumbered for each example in order to avoid unnecessary complications and aggravation. A summary is offered at the end of each section.

Two different mathematical viewpoints are presented: deterministic and stochastic. Analogously to the fact that discussions on the existence of the mathematical reality are reminiscent of discussions about God, discussions about preferring the deterministic or stochastic point of view are reminiscent of disputes between Catholics and Protestants. It is exclusively a question of choice of a mathematical model and nothing more. There are a lot of real problems that can be identically and successfully solved by either a deterministic or a stochastic method.

The ability of an engineer or a geoscientist to generalize or to adapt an existing mathematical approach to a concrete practical problem is very important. Chapter 4 presents examples of mathematical modeling based on generalizing and adapting some of the approaches discussed in Chap. 3.

Chapter 5 illustrates some numerical procedures for the examples discussed. In the conclusion, the fun puzzle on the number of corners of a four-dimensional cube is solved from a mathematical point of view.

Hersh (1997) noted that in a math class everybody gets the same answer, and this indicates a special feature of mathematics: there are right answers. Not right because someone wants it to be right, but right because they are right. I hope that after reading this book the reader can generalize this claim to "in the math class everybody gets the same answer applying the same mathematical models, the same mathematical spaces." The question of what is "the right answer" is philosophical. There are mathematical spaces where $2 + 2$ is not equal to 4. One can say it contradicts the reality, but this is not the case.

When a mathematician develops a model, he/she sets certain assumptions for it. Without fulfilling these assumptions, an application of the model is not only not allowed, it can be even dangerous. Mathematical methods should not be like a mask covering the incompetence of their users. An engineer or a geoscientist should prove the fulfillment of model assumptions before applying the model. He/she should have the courage to reject a model if it seems to be inconsistent with his/her reality. Finally, he/she should have the courage to generalize the model or to develop a better one. This is an idealistic point of view, but one that is true.

The main focus of this book is a demonstration of the superb elegance and great usability of mathematical methods and the goal is to encourage readers to apply them fairly.

Acknowledgements The author thanks her husband, Dr. Konrad Waelder, for helpful discussions and for a critical review of the manuscript.

Chapter 2
From a Problem to Its Solution
by Mathematical Modeling

Mathematical modeling is the basis of many of the sciences. Application of mathematical models has increased rapidly and has spread to most disciplines, and there is reason to believe that this trend will continue to accelerate in the foreseeable future. But, what exactly is mathematical modeling? And, more important, how does it work?

A "real-world" problem is simplified, dissected, and phrased in a mathematical setting. The "usual reality" is replaced by a "mathematical reality." A mathematical model is an object with enormous potential and plays an essential role in many areas of modern applied science. But a mathematical model is not only an object it is also a dynamic process. Objects are independent of individual consciousness. They have relatively permanent qualities. But mathematical models can vary. Following Hersh's definition (1997), "An object is a slow process. A process is a speedy object." Any phenomenon is seen as an object or a process, depending on the chosen scale. Using the scale of the "mathematical reality," we call mathematical modeling an active dynamic process. Not at least, this dynamics also leads to a competition between mathematicians. Gauss discovered the well-known method of least squares before Legendre and abelian integrals before Abel.

Many developments in mathematics antedate the demand for them in the real world and are often not understood, ignored, or rejected. Thus, it can appear that mathematical modeling is a "frozen" object from the viewpoint of "usual reality." A mathematical model is something "God-given" and unassailable. Outsiders in mathematics are taken in by this myth. Insiders are not.

2.1 What Is a Mathematical Model?

Many applications of specific mathematical models are discussed in scientific books, but the question of the definition of a mathematical model is touched on only by the way of exception. Fowler (1997) wrote of mathematical modeling: "There are no set rules, and an understanding of the "right" way to model can only be reached by familiarity with a wealth of examples. [...] A model is a representation of a process." Fowkes and Mahony (1994) noted: "Mathematics in its own right has been becoming increasingly powerful because of the beautiful abstractions which allow one to concentrate on the essentials of what would otherwise be complex arguments." But how is a model created?

If you are asked to describe your best friend, you might start with anything, but it is inconceivable that you would mention his blood pressure. Your description is a model of your friend. But if your friend were to go to a physician, the doctor would probably be interested in his blood pressure. The doctor's examination will create *another* model of your friend.

Again according to Fowler (1997):

> Applied mathematicians have a procedure, almost a philosophy, that they apply when building models. First, there is a phenomenon of interest that one wants to describe or, more importantly, explain. Observations of the phenomenon lead, sometimes after a great deal of effort, to a hypothetical mechanism that can explain the phenomenon. The purpose of a model is then to formulate a description of the mechanism in quantitative terms, and the analysis of the resulting model leads to results that can be tested against the observations.

A model is a simplification. It can be true; it can be incomplete; it can be false. Moreover, a model is an idealization and is always limited in its applicability. One starts with the simplest model, which can be made more complicated. If a model or its generalization does not suit the purpose, it should be rejected and replaced by another. Mathematical modeling is a process.

2.2 Choosing or Deriving an "Optimal" Model

A mathematical problem begins with the identification of a problem—of a phenomenon. Something requires an explanation—a description. There is something we do not understand. Or there something we are missing and want to obtain.

The word "research" has its origin in a term that means "to go around," "to explore," derived from the word meaning "circle." Plutchik (1968) wrote: "The process of searching, observing, and describing has sometimes been called 'naturalistic observation' and has been thought of as rather a primitive kind of research procedure." Owing to important improvements in modern techniques, more accurate observations can be obtained. Although this is often the case, different researchers can still observe different effects.

In creating a mathematical model our first step is to define variables and to develop constitutive relations among them. We substantiate each of them by empirical reasoning. Be that as it may, the process requires a lot of intuition. Hersh (1997) noted that intuition is not direct perception of something external, but rather an effect of the mind/brain manipulating concrete objects—at a later stage of making marks on paper, and still later of manipulating mental images. There is no common rule for triggering a scientist's intuition. But there are ways to train a modeler to recognize the most important variables and relationships.

We begin with a rough, simple model and test it. The advantages and limitations of our primary model should be clear. By the second step, we add one more variable or replace a linear relation between the variables by a nonlinear one. Such generalizations should be substantiated. The aim is to get step by step closer to the mechanism of the real phenomenon.

Once a problem is identified and a mechanism proposed in a "human language," we have to formulate it mathematically. Here, there are several possible ways: One individual prefers a simple model; another wants a high level of complexity. There is no unique "right answer." Different modelers obtain different models. Engineers and geoscientists tend toward the view that a simple, speedily and robustly worked model should be preferred. Geoscientists are often confronted with the fact that many phenomena in nature, such as valleys, cannot be "exactly" defined in a mathematical sense: where exactly does a cliff end and a valley begin? Mathematicians want to establish the robustness of a model, owing to the view that such results help to design and to validate suitable numerical solution procedures (Fowler 1997).

Some restrictions should be made in any case. Often a model is numerical, so its solution has to be numerical as well. The computation process sets limits on the applicability of a model. Fowler (1997) wrote that it is advisable to make a model dimensionless, for it is then possible to determine whether different terms are large or small in a rational way. Neglecting various small terms leads to simplification and some insignificant variables or relations can be ignored.

Every modeler is confronted with the following dilemma: a complex model with an expanding number of variables and relations seems to be close to the reality, but is often not realizable from a numerical point of view. Briefly, a model can be considered as "optimal," if it suits its purpose and can be implemented.

The foregoing was about mathematical modeling from the point of view of a researcher. But I would add some remarks from a psychological point of view. How does one really derive a model? Some scientists only generate their own ideas, ignore existing approaches, and finally develop a wheel after a great deal of effort. Sometimes it can even be amazing and is not absolutely false. Hersh (1997) wrote: "One of my honored teachers (a world-class mathematician) astonished me by saying he didn't read mathematical papers. When something interesting happens, somebody tells him. The same was reported of David Hilbert, in his day the world champion of mathematics. He didn't read."

There are also scientists who start with an in-depth literature search long before a phenomenon is identified. But, if you prefer to go through this thick wood, then try to come back with a solution!

To find a happy medium is optimal.

2.3 Is a Model Good or Bad: Some Arguments for Discussing Results

We assume that a mathematical model is chosen and applied. An engineer or a geoscientist has to comment on the final results. Should the chosen model be accepted or rejected? Fowler (1997) wrote about model validation:

> Ideally, a mathematical model ends by returning to its origin. We look to see whether the model and its analysis explain the phenomenon we are interested in. Does the predicted curve fit the experimental data? Does the predicted stability curve agree with the experimentally determined values? The whole art of mathematical modelling lies in its self-consistency. Science is inaccurate if it derives its justification from the fact that apparently arbitrary assumptions seem to work. And ultimately, the justification of a model is given by the following: It helps us to understand an experimental observation. There is no unique or "correct" model; but there are good models and bad models. The skill of modelling lies in being able to judge which statement holds.

This is the point of view of a mathematician. He/she is familiar with special methods to judge the quality of mathematical models in his/her mathematical reality. But an engineer or a geoscientist may not be able to do so. Is there another way for model validation beyond this "pure mathematical" one? Really, it is possible to compare a Ford with a Mercedes without a degree in engineering. It is a matter of sorting through arguments for such discussions.

We discuss some possible "measures of goodness" for the mathematical models presented in this book.

Chapter 3
Some Real Problems and Their Solutions

Obviously, it is not possible to discuss all the real problems in this one book, but we have included a range of practical problems that should be of interest to engineers and geoscientists. When they have to deal with the kind of problem presented here, engineers and geoscientists often begin by leafing through a pile of mathematics books. Our aim in this volume is explain the basic idea of a mathematical approach in a comprehensible way and gently prepare our readers so that they can access the specialized "hard" mathematics literature relevant to their particular topic. We offer discussions concerning the advantages and disadvantages of certain approaches and comment on the similarities among them. At the end of each section we present a summary of the approaches that have been dealt with and mention the relevant "hard" literature for interested readers. We try throughout to remain true to the central theme of this book: the presentation of the superb elegance, the extensive usability, and the dynamic of mathematical modeling.

3.1 Prediction of a Value: Creating, Refining, or Changing Measurement Grids

Every analysis of spatial and temporal structures starts with data sampling and data description, and the first errors can occur as early as during data sampling. These errors can be roughly divided in two classes: systematic and random. The intrusion of the systematic errors can often be avoided by using modern measurement technologies or by replacing unqualified staff. The influence of random errors can be overcome by repeated measurements of the same type—the main argument for the justification of the methods used in classical statistics.

But what happens if we have only *a single* occurrence of a spatial and temporal phenomenon? For example, we obtain temporal measurements of a process and want to find a time-dependent structure? Or we have a table with results from some irregularly spaced oil-boreholes and have to produce a complete map of the oil reserves? The first problem leads to data fitting by a functional relation, whereas the second results in the creation of a grid of data at nonobserved locations. Solving these problems requires different kinds of predictions.

The first step consists of a description and a preanalysis of the data. The data sets have to be checked for outlier values, as outliers can perceptibly distort the structure of the characteristics that are used in mathematical modeling methods. Detecting and filtering outliers constitutes a special field in mathematics. Mapping the position of samples, plotting temporal data representations, and histograms can help to obtain a first intuitive impression about the data structure. In the following, we assume that the data sets are still free of outliers.

Now the data structure has to be interpreted. All of the causal relations among the observed variables and all the deterministic spatial and temporal structures have to be identified. This interpretation step has an enormous influence on model building.

Based partly on intuition, partly on a priori information, and partly on our experience, we set a final frame, a space for our mathematical modeling. After we set the mathematical model the real work can start.

3.1.1 Deterministic Point of View: Interpolation and Approximation Methods

A deterministic point of view means modeling without considering randomness. Briefly summarized, *inter*polation methods describe the prediction of a value *within* the definition area of the sample points. *Extra*polation methods deal with estimation *beyond* this area. This is true for the prediction of a single value as well as for the global prediction of values by a functional relation at any point in the given area. Moreover, this functional relation should follow the so-called *interpolation demand*, which means that the prediction method returns the true z-values (observed measurements) at the given points (data nodes, data locations). If this assumption is not fulfilled the functional relation leads to a smooth curve (or surface) going through the set of observed values.

Interpolation (and extrapolation) by a functional relation is similar to *approximation*. Most approximation methods replace a given object (discrete sample points, a complex function) by a more-or-less smooth and simple function. The approximation domain and the way in which the data points are distributed in this domain may influence the choice of approximation method. For example, if the data values are specified at the nodes of a regular mesh (grid data) in the approximation domain, there are specific approaches that can exploit this situation.

Reasons and objectives for curve and surface fitting vary: for example, functional representation, data prediction, data smoothing, and data reduction. If we assume that the functional form of an approximating functional relation is more or less immaterial, then piecewise polynomials, as the simplest objects in mathematical modeling, should be considered. .

In this context we roughly distinguish two classes of problems:

1. A single z-value has to be estimated (for mapping, for constructing grids, for reconstructing a missing value).
2. A global prediction by a functional relation has to be made (for data reduction, for generalization, for recognizing structures, for comparison of data sets, for a detailed analysis of data structures).

For these classes, the following refining into subclasses by an initial data specification is possible: (A) Data points are specified at the nodes of a regular mesh. (B) Data points are irregularly, more-or-less chaotically placed.

Different roads lead not only to Rome, but also to solution of the problems that we denote with 1.A, 1.B, 2.A, and 2.B. Here, we present the most familiar methods for one-dimensional (1D) and two-dimensional cases (2D) separately. The presentation of each approach starts with a simple example for better understanding its basic idea. Further, we formulate the common rule. We denote the data points (data locations, variables) by x (1D) or x,y (2D) and the data values (measurements) by z.

Problem 1.A. *Prediction of the z-value at a single point for regularly spaced data (measurements distributed over a mesh, regular time series).*

Most estimation approaches are based on the same idea: a special weighting of known z-values and their linear sum. We start with an explanation of this basic idea by discussing the so-called *generalized arithmetical mean*.

3.1.1.1 Generalized Arithmetical Mean, 1D

Example 3.1.1.1 We consider the following temporal measurements:

$$z_1 = z(1) = 0.1, \ z_2 = z(4) = 0.2, \ z_3 = z(7) = -0.1, \ z_4 = z(10) = -0.2$$

How can the value $z_0 = z(x_0)$ at point $P_0 = x_0 = 1.3$ be estimated? Using basic school knowledge we can get the mean of the z-values at its neighboring points $x = 1$ and $x = 4$. Obviously, this basic arithmetical mean leads to $z_0 = z(1.3) = 0.15$. Here, both measurements $z = 0.1$ and $z = 0.2$ are identically weighted. In other words, the basic arithmetical mean of two z-values can be represented as the proportion of the sum of measurements identically weighted with weight 1 to the sum of these weights:

$$z_0 = z(x_0) = \frac{(z_1 \cdot 1 + z_2 \cdot 1)}{(1+1)} = \frac{0.1 + 0.2}{2} = 0.15 \tag{*.1}$$

Thanks to the engineer's intuition, we note that the point $P_0 = x_0 = 1.3$ is placed *nearer* to $x = 1$ than to $x = 4$. Thus, it would be logical to propose that the z-value at the point $x = 1$ influences the estimation of the z-value at point $x = 1.3$ *more strongly* than the z-value at the point $x = 4$. But now we need a measure to describe this influence objectively. Without any additional information about the *real* nature of this influence, we can consider the inverse distance from the point $x = 1.3$ to both neighbors $x = 1$ and $x = 4$ for constructing weights. The influence of known z-values on the z-value, which has to be predicted, becomes stronger with decreasing distance between the data and the prediction point. Therefore, the generalized arithmetical mean of two z-values corresponds to:

$$z_0 = z(x_0) = \frac{\left(z_1 \cdot \dfrac{1}{0.3} + z_2 \cdot \dfrac{1}{2.7}\right)}{\left(\dfrac{1}{0.3} + \dfrac{1}{2.7}\right)} = \frac{(z_1 \cdot 2.7 + z_2 \cdot 0.3)}{(2.7 + 0.3)} = \frac{0.1 \cdot 2.7 + 0.2 \cdot 0.3}{3} = 0.11 \quad (*.2)$$

This estimation method is also called *one-dimensional linear interpolation—* "linear" because the spatial distribution of all predicted z-values between $x = 1$ and $x = 4$ follows a line from point $(x_1, z_1) = (1, 0.1)$ to point $(x_2, z_2) = (4, 0.2)$.

But now what happens if *all* given z-values (not only the nearest neighbors of the prediction point) are considered to estimate the z-value at P_0? No problem: the basic idea of inverse-distance-dependent weighting remains unchanged:

$$z_0 = z(x_0) = \frac{\left(z_1 \cdot \dfrac{1}{0.3} + z_2 \cdot \dfrac{1}{2.7} + z_3 \cdot \dfrac{1}{5.7} + z_4 \cdot \dfrac{1}{8.7}\right)}{\left(\dfrac{1}{0.3} + \dfrac{1}{2.7} + \dfrac{1}{5.7} + \dfrac{1}{8.7}\right)}$$

$$= \frac{\left(\dfrac{0.1}{0.3} + \dfrac{0.2}{2.7} + \dfrac{-0.1}{5.7} + \dfrac{-0.2}{8.7}\right)}{\left(\dfrac{1}{0.3} + \dfrac{1}{2.7} + \dfrac{1}{5.7} + \dfrac{1}{8.7}\right)} = 0.0919$$

$$(*.3)$$

Finally, we may limit the maximal distance between a data point and the point of prediction where the influence still exists based upon an engineer's or a geoscientist's experience and practical knowledge. In this example, we assume that this maximal distance should be equal to 6. Thus, we get an estimation based on only three out of four given z-values as

$$z_0 = z(x_0) = \frac{\left(z_1 \cdot \dfrac{1}{0.3} + z_2 \cdot \dfrac{1}{2.7} + z_3 \cdot \dfrac{1}{5.7}\right)}{\left(\dfrac{1}{0.3} + \dfrac{1}{2.7} + \dfrac{1}{5.7}\right)} = \frac{\left(\dfrac{0.1}{0.3} + \dfrac{0.2}{2.7} + \dfrac{-0.1}{5.7}\right)}{\left(\dfrac{1}{0.3} + \dfrac{1}{2.7} + \dfrac{1}{5.7}\right)} = 0.1005$$

$$(*.4)$$

Moreover, we can use an *influence function* constructed based on experience or practical knowledge instead of depending on inverse distances for determining weights. Furthermore, the influence of sample points may increase for certain distances and be constant for others. The well-known *method of inverse squares* deals with an influence function in the following form:

$$f(d) = \frac{1}{d^2}, \; d = |x - x_0| \qquad\qquad (*.5)$$

where d is the distance between a data point x and the point of prediction x_0. The concrete choice of an influence function depends on the practical application.

The solutions (*.1)–(*.4) are obviously different from one another, but without further information we do not know the best result. Applying different mathematical models for the same problem can lead to different results.

Now we formulate *the common rule for using the generalized arithmetical mean in the 1D-case:*

Let $x_1, x_2, \ldots, x_N$ be data points (locations) and $z_1, z_2, \ldots, z_N$ be z-values (measurements) at these points. A continuous function $f(d)$ defined for $d \in (o, \infty)$ is called an influence function. The prediction of the z-value at any point x_0 yields

$$z(x_0) = \begin{cases} z(x_i) = z_i, \; if \; x_0 = x_i \; or \; d_i = |x_i - x_0| = 0, \quad i = 1 \ldots N \\[2mm] \dfrac{\sum\limits_{i=1}^{N} z_i \cdot f(d_i)}{\sum\limits_{i=1}^{N} f(d_i)}, \; if \; d_1 > 0 \ldots d_N > 0 \end{cases} \qquad (3\text{-}1)$$

The simplest forms of the function $f(d)$ are

$$f(d) = \frac{1}{d} \; and \; f(d) = \frac{1}{d^2}$$

3.1.1.2 Generalized Arithmetical Mean, 2D

The basic idea is similar to that in the 1D-case, namely the influence of the data points is assumed to be dependent on the distance to the point of prediction. The calculation of the distance between two points $P_0 = (x_0, y_0)$ and $P_i = (x_i, y_i)$ in the plane is obviously equal to $d_i = \sqrt{(x_i - x_0)^2 + (y_i - y_0)^2}, \; i = 1 \ldots N$.

Example 3.1.1.2 We consider the following measurements:

$$z_{11} = z(P_1) = z(1,0) = 0.1, \; z_{21} = z(P_2) = z(4,0) = 0.2,$$
$$z_{12} = z(P_3) = z(1,1) = -0.1, \; z_{22} = z(P_4) = z(4,1) = -0.2.$$

We have to estimate the z-value at the point $P_0 = (x_0, y_0) = (2, 0.5)$ using all four sample points. At first, we calculate the distance from each data point to the point of prediction: $d_1 = d_3 = 1.1180$, $d_2 = d_4 = 2.0616$.

Now we use the influence function $f(d) = 1/d^2$ in order to predict the z-value at point $P_0 = (x_0, y_0) = (2, 0.5)$:

$$z_0 = z(P_0) = \frac{\left(z_1 \cdot \dfrac{1}{1.1180^2} + z_2 \cdot \dfrac{1}{2.0616^2} + z_3 \cdot \dfrac{1}{1.1180^2} + z_3 \cdot \dfrac{1}{2.0616^2}\right)}{\left(2 \cdot \dfrac{1}{1.1180^2} + 2 \cdot \dfrac{1}{2.0616^2}\right)}$$

$$= \frac{\left(\dfrac{0.1}{1.1180^2} + \dfrac{0.2}{2.0616^2} + \dfrac{-0.1}{1.1180^2} + \dfrac{-0.2}{2.0616^2}\right)}{\left(2 \cdot \dfrac{1}{1.1180^2} + 2 \cdot \dfrac{1}{2.0616^2}\right)} = 0$$

$$(*.1)$$

Certainly, other influence functions are possible. By the way, we can predict the z-value at any point in the plane, not only within the given four measurements. However, is doubtful from a practical point of view if these four data points can affect the z-value located, for example, at $P_0 = (x_0, y_0) = (200, 500)$, so such extrapolations should be avoided.

We are allowed to consider additional data points—if such points are given—beyond the grid cell including the point of estimation. The maximal distance at which to describe the influence on the interesting point of the prediction should be chosen intuitively or from a practical view.

There is also an analogous method often called *bilinear interpolation for the 2D-case*, but this method only works within a grid cell including the point of estimation. The basic idea is similar to the linear interpolation for the 1D-case. Here, however, the influence of the z-values (measurements) is determined by replacing the distance between a node of a cell and the point of estimation with areas A_1, A_2, A_3, A_4 of squares S_1, S_2, S_3, S_4, as shown in Fig. 3.1. The estimated value at point $P_0 = (x_0, y_0) = (2, 0.5)$ is obtained by

$$z_0 = z(P_0) = \frac{\left(z_1 \cdot \dfrac{1}{A_1} + z_2 \cdot \dfrac{1}{A_2} + z_3 \cdot \dfrac{1}{A_3} + z_4 \cdot \dfrac{1}{A_4}\right)}{\left(\dfrac{1}{A_1} + \dfrac{1}{A_2} + \dfrac{1}{A_3} + \dfrac{1}{A_4}\right)}$$

$$= \frac{\left(\dfrac{0.1}{0.5} + \dfrac{0.2}{1} + \dfrac{-0.1}{0.5} + \dfrac{-0.2}{1}\right)}{\left(2 \cdot \dfrac{1}{0.5} + 2 \cdot \dfrac{1}{1}\right)} = 0$$

$$(*.2)$$

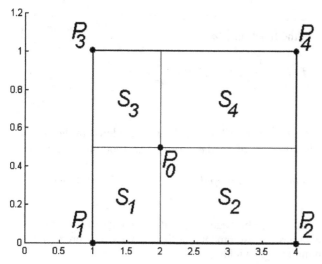

Fig. 3.1 Bilinear interpolation over a square mesh: construction of special weights

Another way is

$$z_0 = z(P_0) = \frac{(z_1 \cdot A_4 + z_2 \cdot A_3 + z_3 \cdot A_2 + z_4 \cdot A_1)}{(A_4 + A_3 + A_2 + A_1)}$$

$$= \frac{(0.1 \cdot 1 + 0.2 \cdot 0.5 + (-0.1) \cdot 1 + (-0.2) \cdot 0.5)}{(2 \cdot 0.5 + 2 \cdot 1)} = 0$$

Here the two methods lead to the same results because of the special data structure. The spatial distribution of the z-values predicted by bilinear interpolation at all points within a grid cell follows a so-called *hyperbolic paraboloid*.

Now we formulate *the common rule for using the generalized arithmetical mean in the 2D-case:*

Let $(x_1, y_1), (x_2, y_2), \ldots, (x_N, y_N)$ be data points (locations) and $z_1, z_2, \ldots, z_N$ be the z-values (measurements) at these points. A continuous function $f(d)$ defined with $d \in (o, \infty)$ is called the influence function. The prediction of the z-value at any point (x_0, y_0) yields

$$z(x_0, y_0) = \begin{cases} z(x_i, y_i) = z_i, & \text{if } P_0 = P_i \text{ or } d_i = \sqrt{(x_i - x_0)^2 + (y_i - y_0)^2} = 0, \\ \dfrac{\sum\limits_{i=1}^{N} z_i \cdot f(d_i)}{\sum\limits_{i=1}^{N} f(d_i)}, & \text{if } d_1 > 0 \ldots d_N > 0, \quad i = 1 \ldots N \end{cases}$$
(3-2)

The simplest forms of the function $f(d)$ are

$$f(d) = \frac{1}{d} \quad and \quad f(d) = \frac{1}{d^2}$$

If *only* the four nodes P_1, P_2, P_3, and P_4 of a cell are considered in the z-value estimation at point P_0 within this cell of the data grid, the method of bilinear interpolation can be used:

The common rule for using bilinear interpolation (2D-case) is as follows:

Let $(x_1, y_1), (x_2, y_2), \ldots, (x_4, y_4)$ be nodes of a cell of a data grid (locations) and $z_1, z_2, \ldots, z_4$ be z-values (measurements) at these points. The prediction of the z-value at any point (x_0, y_0) yields

$$z(x_0, y_0) = \begin{cases} z_i, \ if \ (x_0, y_0) = (x_i, y_i) \ or \ A_i = |(x_i - x_0) \cdot (y_i - y_0)| = 0, \quad i = 1 \ldots 4 \\ \dfrac{\sum\limits_{i=1}^{4} z_i \cdot \dfrac{1}{|(x_i - x_0) \cdot (y_i - y_0)|}}{\sum\limits_{i=1}^{4} \dfrac{1}{|(x_i - x_0) \cdot (y_i - y_0)|}}, \ if \ A_1 > 0, \ldots, A_4 > 0, \end{cases}$$

(3-3)

Problem 1.B. *Prediction of the z-value at a single point has to be done for irregularly spaced data (chaotically distributed measurements).*

The first approach is the use of the generalized arithmetical mean as given for Problem 1.A., and this seems to be the most relevant method for the 1D-case.

For the 2D-case there is a second possibility, which starts with the so-called *triangulation* of sample points. This means that the data points are joined by lines in such a way that a complete covering of the defined area of measurements is achieved. This covering consists solely of triangles; the triangulation procedure is not trivial, and certain conditions have to be fulfilled in order to get an "optimal" triangular network or triangular mesh. Currently, this procedure is available in most software tools for engineers and geoscientists that deal with spatial or geographic data. After triangulation, the predicted value can be obtained following a principle similar to the one described for bilinear interpolation.

Example 3.1.1.3 We consider a single triangle from a triangular mesh including the prediction point $P_0 = (0.5, 0.3)$ with following measurements:

$$z_1 = z(P_1) = z(0.0, 0.0) = 0.1, \quad z_2 = z(P_2) = z(1.0, 0.0) = 0.2$$

and

$$z_3 = z(P_3) = z(0.6, 0.9) = -0.5.$$

Analogously to the bilinear interpolation for a square mesh, the estimation of the z-value at point $P_0 = (0.5, 0.3)$ within the triangle $P_1P_2P_3$ considers the areas A^1, A^2, A^3 of the corresponding sets S^1, S^2, S^3. Each set is placed opposite the point with the same number (see Fig. 3.2), so it may be assumed that the influence of given z-values on the estimation of the unknown z-value at point P_0 increases with expanding corresponding areas. Thus, we obtain

$$z_0 = z(P_0) = \frac{(z_1 \cdot A^1 + z_2 \cdot A^2 + z_3 \cdot A^3)}{(A^1 + A^2 + A^3)}$$

$$= \frac{(0.1 \cdot 0.165 + 0.2 \cdot 0.135 + (-0.5) \cdot 0.15)}{0.45} = -0.0315$$

$$(*.1)$$

Here we can apply the following very useful equation for calculating the triangle area based on the so-called vector product:

$$A^1 = \frac{1}{2} \cdot |P_2P_3 \times P_2P_0| = \frac{1}{2} \cdot \begin{vmatrix} i & j & k \\ x_3 - x_2 & y_3 - y_2 & 0 \\ x_0 - x_2 & y_0 - y_2 & 0 \end{vmatrix}$$

$$= \frac{1}{2} \cdot |(x_3 - x_2)(y_0 - y_2) - (x_0 - x_2)(y_3 - y_2)| \qquad (*.2)$$

$$= \frac{1}{2} \cdot |(0.6 - 1.0)(0.3 - 0.0) - (0.5 - 1.0)(0.9 - 0.0)| = 0.165,$$

$$A^2 = \frac{1}{2} \cdot |P_1P_3 \times P_1P_0| = 0.135, \quad A^3 = \frac{1}{2} \cdot |P_1P_2 \times P_1P_0| = 0.15$$

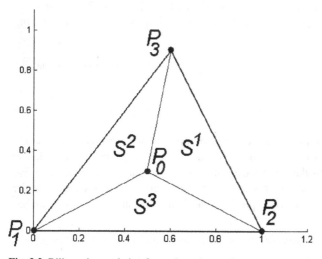

Fig. 3.2 Bilinear interpolation for a triangular mesh: construction of special weights

Problem 2.A. *Regular data (measurements on a mesh/grid) should be interpolated (or described) applying a functional relation.*

There are many solution methods available for this problem. Sets of regular data (data grids) are more "comfortable"—especially with respect to the viewpoint of a software developer—for approximation by a functional relation compared with irregular data sets. Here we want to distinguish two classes of methods. The first class includes approaches for which the interpolation demand has to be fulfilled (Problem 2.A-D). The methods from the second class renounce this demand (Problem 2A-ND).

Problem 2.A-D *Regular data (measurements on a mesh, a grid) should be interpolated (be described) applying a functional relation with exactly reproduced measurements.*

Again, at first the "good old" generalized arithmetical mean described above can be used. We replace the concrete coordinates x_0 (1D-case) and (x_0, y_0) (2D-case) of the single point P_0, where the z-value has to be predicted, by "free" coordinates x for the 1D-case and (x, y) for the 2D-case. With this replacement and the use of (3-1) and (3-2), the necessary functional relation based on the generalized arithmetical mean can be easily obtained. This method is popular in geodetic applications. Chapter 5 includes examples of code for implementation.

The alternative method goes back on the great French mathematician Lagrange and is called *Lagrange's interpolation method.*

3.1.1.3 Lagrange's Interpolation Method

We start with the same trivial data sets as in Examples 3.1.1.1 and 3.1.1.2 and distinguish the 1D-case and 2D-case for better understanding as well as for comparing results.

Example 3.1.1.1' (1D-Case) Again, we consider the following temporal measurements:

$$z_1 = z(1) = 0.1,\ z_2 = z(4) = 0.2,\ z_3 = z(7) = -0.1,\ z_4 = z(10) = -0.2$$

What functional relation describes these data points so that the z-values obtained at given points agree with the estimation? The basic idea of Lagrange's interpolation method is also based on the generalized arithmetical mean explained earlier. We have to set proper *variable weights* for the given z-value in such a way that the known z-values would be exactly reproduced at their location points. Thus, we are looking for weights that

$$z(x) = z_1 \cdot w_1(x) + \ldots + z_4 \cdot w_4(x) = \sum_{i=1}^{4} z_i \cdot w_i(x)$$
$$= 0.1 \cdot w_1(x) + 0.2 \cdot w_2(x) + (-0.1) \cdot w_3(x) + (-0.2) \cdot w_4(x), \qquad (*.1)$$

$$w_i(x) = \begin{cases} 1, & x = x_i \\ 0, & x = x_j, \ j \neq i, \ i,j = 1 \ldots 4 \end{cases}$$

Which weights could fulfill this condition? Let us test the following weights constructed according to Lagrange's concept:

$$w_1(x) = \frac{(x - x_2)(x - x_3)(x - x_4)}{(x_1 - x_2)(x_1 - x_3)(x_1 - x_4)} = \frac{(x-4)(x-7)(x-10)}{(1-4)(1-7)(1-10)},$$
$$(*.2)$$
$$w_2(x) = \frac{(x - x_1)(x - x_3)(x - x_4)}{(x_2 - x_1)(x_2 - x_3)(x_2 - x_4)} = \frac{(x-1)(x-7)(x-10)}{(4-1)(4-7)(4-10)},$$

$$w_3(x) = \frac{(x - x_1)(x - x_2)(x - x_4)}{(x_3 - x_1)(x_3 - x_2)(x_3 - x_4)} = \frac{(x-1)(x-4)(x-10)}{(7-1)(7-4)(7-10)},$$
$$(*.3)$$
$$w_4(x) = \frac{(x - x_1)(x - x_2)(x - x_3)}{(x_4 - x_1)(x_4 - x_2)(x_4 - x_3)} = \frac{(x-1)(x-4)(x-7)}{(10-1)(10-4)(10-7)}.$$

It can be seen that we get the weight $w_1(x_1) = 1$ in (*.2) by replacing the variable x by $x = x_1 = 1$. At the same time the other weights in (*.2) and (*.3) are equal to zero. Using these weights we get

$$z(x_1) = 0.1 \cdot 1 + 0.2 \cdot 0 + (-0.1) \cdot 0 + (-0.2) \cdot 0 = 0.1 = z_1, \qquad (*.4)$$

which indicates that the interpolation demand at the point $x = x_1 = 1$ has been fulfilled. Analogously, after replacing the variable x by $x = x_2 = 4$, $x = x_3 = 7$, $x = x_4 = 10$, we get

$$w_2(x_2) = w_2(4) = 1, \ w_1(4) = w_3(4) = w_4(4) = 0 \Rightarrow z(x_2) = z_2 = 0.2,$$
$$w_3(x_3) = w_3(7) = 1, \ w_1(7) = w_2(7) = w_4(7) = 0 \Rightarrow z(x_3) = z_3 = -0.1,$$
$$w_4(x_4) = w_4(10) = 1, \ w_1(10) = w_2(10) = w_3(10) = 0 \Rightarrow z(x_4) = z_4 = -0.2$$
$$(*.5)$$

Summarizing, we have

$$z(x) = 0.1 \cdot \frac{(x-4)(x-7)(x-10)}{(1-4)(1-7)(1-10)} + 0.2 \frac{(x-1)(x-7)(x-10)}{(4-1)(4-7)(4-10)}$$
$$(*.6)$$
$$+ (-0.1) \cdot \frac{(x-1)(x-4)(x-10)}{(7-1)(7-4)(7-10)} + (-0.2) \cdot \frac{(x-1)(x-4)(x-7)}{(10-1)(10-4)(10-7)}$$

The function (*.6) is the functional relation we are looking for. An interpolation polynomial of degree three (see Fig. 3.3), it describes the structure of the given data points so that the z-values obtained at these points are reproduced by the

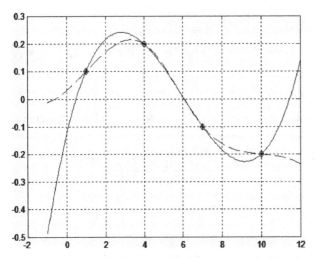

Fig. 3.3 An interpolating polynomial of degree three (*bold*) and a cubic spline (*dashed*) for the data set from Example 3.1.1.1

estimation. By using any other x-value the corresponding z-value can be predicted. This polynomial is only an exemplary polynomial of degree three that fits this data set and simultaneously fulfills the interpolation demand. Of course, there are other analytical functions that fulfill this condition.

Now, *the common rule for Lagrange's interpolation method (1D-case)* can be formulated:

Let $x_1, x_2, \ldots, x_N$ be data points (locations) and $z_1, z_2, \ldots, z_N$ be the z-values (measurements) at these points. A polynomial fitting these data so that the interpolation demand is fulfilled has the following form:

$$z(x) = \sum_{i=1}^{N} z_i \cdot w_i(x),$$

$$w_i(x) = \frac{(x-x_1)(x-x_2)\ldots(x-x_{i-1})(x-x_{i+1})\ldots(x-x_N)}{(x_i-x_1)(x_i-x_2)\ldots(x_i-x_{i-1})(x_i-x_{i+1})\ldots(x_i-x_N)}, \quad i=1,\ldots,N$$

$$w_i(x_j) = \delta_{ij} = \begin{cases} 1, & j=i \\ 0, & j\neq i \end{cases}$$

$$(3\text{-}4)$$

Example 3.1.1.2' *(2D-Case)* Again, we consider the following measurements:

$$z_{11} = z(P_1) = z(1,0) = 0.1, \quad z_{21} = z(P_2) = z(4,0) = 0.2,$$
$$z_{12} = z(P_3) = z(1,1) = -0.1, \quad z_{22} = z(P_4) = z(4,1) = -0.2.$$

Following the 1D-case, we are asking for a functional relation that fits the data set in such a way that the interpolation demand is fulfilled. Thus, the given measured z-values have to be reproduced exactly. For simplification of further equations, we number the z-values and points with double indices 11, 21, 12, and 22 corresponding to the ordered numbering of x- and y-values in the plane. The special weighting of measured z-values is also used here:

$$z(x,y) = z_{11} \cdot w_{11}(x,y) + z_{12} \cdot w_{12}(x,y) + z_{21} \cdot w_{21}(x,y) + z_{22} \cdot w_{22}(x,y) = \sum_{i,j=1}^{2} z_{ij} \cdot w_{ij}(x,y)$$

$$= 0.1 \cdot w_{11}(x,y) + 0.2 \cdot w_{21}(x,y) + (-0.1) \cdot w_{21}(x,y) + (-0.2) \cdot w_{22}(x,y),$$

$$w_{ij}(x) = \begin{cases} 1, & x = x_i, \ y = y_j \\ 0, & x = x_k, \ y = y_l, \ k \neq i, \ l \neq j; \ i,j,k,l = 1,2 \end{cases}$$

$$(*.1)$$

We construct the variable weights by:

$$w_{11}(x,y) = \frac{(x-x_2)(y-y_2)}{(x_1-x_2)(y_1-y_2)} = \frac{(x-4)(y-1)}{(1-4)(0-1)},$$
$$w_{12}(x,y) = \frac{(x-x_2)(y-y_1)}{(x_1-x_2)(y_2-y_1)} = \frac{(x-4)(y-0)}{(1-4)(1-0)},$$

$$(*.2)$$

$$w_{21}(x,y) = \frac{(x-x_1)(y-y_2)}{(x_2-x_1)(y_1-y_2)} = \frac{(x-1)(y-1)}{(4-1)(0-1)},$$
$$w_{22}(x,y) = \frac{(x-x_1)(y-y_1)}{(x_2-x_1)(y_2-y_1)} = \frac{(x-1)(y-0)}{(4-1)(1-0)}.$$

$$(*.3)$$

Analogously to Example 3.1.1.1' it can be seen that the interpolation demand at given points is fulfilled. Summarizing, we have

$$z(x,y) = 0.1 \cdot \frac{(x-4)(y-1)}{(1-4)(0-1)} + 0.2 \cdot \frac{(x-4)(y-0)}{(1-4)(1-0)} + (-0.1) \cdot \frac{(x-1)(y-1)}{(4-1)(0-1)}$$
$$+ (-0.2) \cdot \frac{(x-1)(y-0)}{(4-1)(1-0)}.$$

$$(*.4)$$

Function (*.4) is the functional relation we are looking for. Such two-dimensional polynomials are called *incomplete polynoms* of degree two, because the terms corresponding to x^2 and y^2 are omitted. Further, this is a hyperbolic paraboloid. It describes the structure of the given data points so that the z-values obtained at the given points remain unchanged. Using any other value x,y, we can predict the corresponding z-value. Apart from the proposed function, there are other two-dimensional polynomials suitable for fitting to this data set.

The common rule for Lagrange's interpolation method (2D-case) reads as follows:

Let $x_1, \ldots, x_N, y_1, \ldots, y_M$ be coordinates of data points (locations) and $z_{11}(x_1, y_1) = z_{11}, \ldots, z_{1M}(x_1, y_M) = z_{1M}, \ldots, z_{NM}(x_N, y_M) = z_{NM}$ be the z-values (measurements on a mesh) at these points. The two-dimensional polynomial fitting this data set so that the interpolation demand is fulfilled has the following form:

$$z(x, y) = \sum_{i=1}^{N} \sum_{j=1}^{M} z_{ij} w_{ij}(x, y) \quad with \quad w_{ij}(x, y) = \varphi_i(x) \cdot \varphi_j(y),$$

$$\varphi_i(x) = \frac{(x - x_1)(x - x_2) \ldots (x - x_{i-1})(x - x_{i+1}) \ldots (x - x_N)}{(x_i - x_1)(x_i - x_2) \ldots (x_i - x_{i-1})(x_i - x_{i+1}) \ldots (x_i - x_N)}, \quad i = 1 \ldots N$$

$$\varphi_j(y) = \frac{(y - y_1)(y - y_2) \ldots (y - y_{j-1})(y - y_{j+1}) \ldots (y - y_M)}{(y_j - y_1)(y_j - y_2) \ldots (y_j - y_{j-1})(y_i - y_{j+1}) \ldots (y_j - y_M)}, \quad j = 1 \ldots M$$

$$(3\text{-}5)$$

Remark: Lagrange's interpolation method has advantages and disadvantages. The simple calculation of weights and a proper "stability" corresponding to changing z-values over a fixed (x, y)-grid with calculated weights remaining unchanged are among its advantages. This method can be recommended for investigations based such spatial data sampled on the same grid of coordinates, for example, meteorological data sampled during a temporal interval. The main disadvantage of this method corresponds to the necessity to recalculate the weights completely if the grid is refined by considering additional measurements.

This disadvantage can be removed for the 1D-case by using another interpolation equation, which goes back to the great British scientist Newton.

3.1.1.4 Newton's Interpolation Equation

We go back to Example 3.1.1.1 in order to demonstrate the basic idea of this method.

Example 3.1.1.1″ (1D-Case) We consider the following temporal measurements:

$$z_1 = z(1) = 0.1, \ z_2 = z(4) = 0.2, \ z_3 = z(7) = -0.1, \ z_4 = z(10) = -0.2$$

What is the functional relation describing this data sampling so that the z-values obtained at the measurement points remain unchanged? Moreover, the obtained functional relation should be changed minimally if the original data set is refined by adding a new measurement $z_5 = z(13) = -0.3$. Here we show a recursive approach to obtain such a functional relation. We start with the first measurement and set

$$z(x) = b_1 = z(1) = 0.1 \tag{*.1}$$

This relation still fulfills the interpolation demand at point x_1, but not at the other points. In the second step we refine this relation to

$$z(x) = b_1 + b_2 \cdot (x - x_1) = 0.1 + b_2 \cdot (x - 1) \tag{*.2}$$

It can be seen that this relation still fulfills the interpolation demand at the point x_1 because the second summand (in parentheses) is equal to zero at this point. We choose the parameter b_2 so that $z(4) = 0.1 + b_2 \cdot (4 - 1) = 0.2$ holds, fulfilling the interpolation demand at point x_2. This leads to $b_2 = 1/30$ and $z(x) = 0.1 + (1/30)(x - 1)$, which fulfills the interpolation demand at both points x_1 and x_2, but still not at points x_3 and x_4. We continue refining to get

$$\begin{aligned} z(x) &= b_1 + b_2 \cdot (x - x_1) + b_3 \cdot (x - x_1)(x - x_2) \\ &= 0.1 + \tfrac{1}{30} \cdot (x - 1) + b_3 \cdot (x - 1)(x - 4) \end{aligned} \tag{*.3}$$

Now we calculate b_3 from

$$z(7) = 0.1 + \frac{1}{30} \cdot (7 - 1) + b_3 (7 - 1)(7 - 4) = -0.1 .$$

After this step we have:

$$z(x) = 0.1 + \frac{1}{30} \cdot (x - 1) + \left(-\frac{1}{45}\right) \cdot (x - 1)(x - 4) .$$

In the last step we assume

$$\begin{aligned} z(x) &= b_1 + b_2 \cdot (x - x_1) + b_3 \cdot (x - x_1)(x - x_2) + b_4 \cdot (x - x_1)(x - x_2)(x - x_3) \\ &= 0.1 + \frac{1}{30} \cdot (x - 1) + \left(-\frac{1}{45}\right) \cdot (x - 1)(x - 4) + b_4 \cdot (x - 1)(x - 4)(x - 7) \end{aligned} \tag{*.4}$$

and calculate b_4 from

$$\begin{aligned} z(10) = 0.1 + \tfrac{1}{30} \cdot (10 - 1) + \left(-\tfrac{1}{45}\right) \cdot (10 - 1) \\ (10 - 4) + b_4 \cdot (10 - 1)(10 - 4)(10 - 7) = -0.2 \end{aligned} \tag{*.5}$$

The following functional relation is the one that we are looking for:

$$z(x) = 0.1 + \frac{1}{30} \cdot (x - 1) + \left(-\frac{1}{45}\right) \cdot (x - 1)(x - 4) + \frac{1}{270} \cdot (x - 1)(x - 4)(x - 7) . \tag{*.6}$$

After we simplify, we can easily prove that polynomial (*.6) is the same as the one obtained by Lagrange's interpolation method in Example 3.1.1.1′, and this is due to the fact that there is a unique polynomial of degree three that goes exactly through the four given points in the plane.

In contrast to Lagrange's method, Newton's interpolation method is stable with respect to refining a data grid. For example, if we obtain an additional measurement $z_5 = z(13) = -0.3$, we must continue the above procedure after evaluating (*.6) by setting an additional parameter b_5:

$$z(x) = 0.1 + \frac{1}{30} \cdot (x-1) + \left(-\frac{1}{45}\right) \cdot (x-1)(x-4) + \frac{1}{270} \cdot (x-1)(x-4)(x-7)$$
$$+ b_5 \cdot (x-1)(x-4)(x-7)(x-10)$$

$$(*.7)$$

The interested reader can easily calculate this parameter from

$$z(13) = 0.1 + \frac{1}{30} \cdot (13-1) + \left(-\frac{1}{45}\right) \cdot (13-1)(13-4)$$
$$+ \frac{1}{270} \cdot (13-1)(13-4)(13-7) + b_5 \cdot (13-1)(13-4)(13-7)(13-10) = -0.3$$
$$(*.8)$$

Remark leading to the next approach: Lagrange's and Newton's interpolation methods are based on polynomials. Obviously, the degree of these polynomials increases for expanding data sets. This effect logically occurs because the interpolation demand has to be fulfilled. Interpolation polynomials of high degree have the undesirable tendency to oscillate a great deal between the measurement points, so the degree of an interpolation polynomial should be reduced to avoid such oscillation. Using so-called *splines* offers possible solution for this dilemma. Splines are piecewise continuous functions with special smoothing and oscillation-damping properties.

We start by explaining the basic idea of so-called *cubic splines* in the 1D-case.

3.1.1.5 Cubic Splines (1D-Case)

The basic idea of classical splines in the 1D-case corresponds to the *linear interpolation of its second derivatives* at the measurement locations. We start with an example and formulate the common rule for continuous 1D-splines with piecewise cubic functions later. A *cubic function* is an alternative designation for a polynomial of degree three.

Example 3.1.1.1‴ (1D-Case) We return to our well-known temporal measurements:

$$z_1 = z(1) = 0.1, \ z_2 = z(4) = 0.2, \ z_3 = z(7) = -0.1, \ z_4 = z(10) = -0.2$$

We search a *group* of piecewise cubic functions. Each function exists only in an interval between two adjoining locations of measurements. The passage from one function to another should be twice differentiable.

Thus, we start with a group of functions $S_1(x), S_2(x), S_3(x)$ that mathematicians use to denote a "piecewise determined function" $S(x)$, called the *spline function* or the *spline*. We have four measurements, so there are three adjoining intervals:

$$S(x) = \begin{cases} S_1(x), & x \in [1,4], \\ S_2(x), & x \in [4,7], \\ S_3(x), & x \in [7,10] \end{cases} \tag{*.1}$$

We denote the unknown second derivatives at the location points by M_1, M_2, M_3, M_4, assuming that these derivatives exist and these values have to be evaluated in order to construct the corresponding spline. As the passages are twice differentiable,

$$\frac{d^2}{dx^2} S_1(4) = \frac{d^2}{dx^2} S_2(4) = M_2, \quad \frac{d^2}{dx^2} S_2(7) = \frac{d^2}{dx^2} S_3(7) = M_3 \tag{*.2}$$

must hold.

We set the first and the last second derivative (outer second derivatives) equal to zero: $M_1 = M_4 = 0$. This is done because the number of equations has to be equal to the number of variables that have to be determined. If other values of the outer second derivatives are known, for example, from a practical view, they can be used instead of the zeros.

Further, the second derivative at point x between the measurement locations is obtained by the following linear interpolation of appointed second derivatives (see Example 3.1.1.1):

$$\frac{d^2}{dx^2} S(x) = \begin{cases} 0 \cdot \dfrac{(x_2-x)}{(x_2-x_1)} + M_2 \cdot \dfrac{(x-x_1)}{(x_2-x_1)} = M_2 \cdot \dfrac{(x-1)}{(4-1)}, & x \in [1,4], \\ M_2 \cdot \dfrac{(x_3-x)}{(x_3-x_2)} + M_3 \cdot \dfrac{(x-x_2)}{(x_3-x_2)} = M_2 \cdot \dfrac{(7-x)}{(7-4)} + M_3 \cdot \dfrac{(x-4)}{(7-4)}, & x \in [4,7], \\ M_3 \cdot \dfrac{(x_4-x)}{(x_4-x_3)} + 0 \cdot \dfrac{(x-x_3)}{(x_4-x_3)} = M_3 \cdot \dfrac{(10-x)}{(10-7)}, & x \in [7,10] \end{cases}$$

$$\tag{*.3}$$

By setting $x_1 = 1, x_2 = 4, x_3 = 7, x_4 = 10$, it can be easily proved that this linear interpolation reproduces the "exact values" M_1, M_2, M_3, M_4 of the second derivatives at the passage points. Owing to this interpolation, the unwanted oscillations between measurements can be damped.

Now, we can reconstruct the first derivatives of spline $S(x)$ from the second derivatives by integrating (*.3):

$$\frac{d}{dx}S(x) = \begin{cases} \frac{1}{2}M_2 \cdot \frac{(x-1)^2}{(4-1)} + A_1, & x \in [1,4], \\[2mm] -\frac{1}{2}M_2 \cdot \frac{(7-x)^2}{(7-4)} + \frac{1}{2}M_3 \cdot \frac{(x-4)^2}{(7-4)} + A_2, & x \in [4,7], \\[2mm] -\frac{1}{2}M_3 \cdot \frac{(10-x)^2}{(10-7)} + A_3, & x \in [7,10] \end{cases} \quad (*.4)$$

where A_1, A_2, A_3 are constants. The reader may convince himself or herself that (*.4) is true by redifferentiating it to (*.3). Further, we can reconstruct the spline $S(x)$ using the same principle of integration, but in this case we integrate (*.4) and obtain

$$S(x) = \begin{cases} \frac{1}{6}M_2 \cdot \frac{(x-1)^3}{(4-1)} + A_1(x-1) + B_1, & x \in [1,4], \\[2mm] \frac{1}{6}M_2 \cdot \frac{(7-x)^3}{(7-4)} + \frac{1}{6}M_3 \cdot \frac{(x-4)^3}{(7-4)} + A_2(x-4) + B_2, & x \in [4,7], \\[2mm] \frac{1}{6}M_3 \cdot \frac{(10-x)^3}{(10-7)} + A_3(x-7) + B_3, & x \in [7,10] \end{cases} \quad (*.5)$$

where B_1, B_2, B_3 are constants that would vanish if we differentiate (*.5) in order to get (*.4). All the constants A_1, A_2, A_3 and B_1, B_2, B_3 using M_2, M_3 can be evaluated clearly by considering six additional conditions. The interpolation demand for all four measurement locations offers us these conditions because each "piece" $S_1(x), S_2(x), S_3(x)$ of the spline $S(x)$ produces exactly two equations:

$$\begin{array}{ll} S_1(1) = z(1) = 0.1, & S_1(4) = z(4) = 0.2, \\ S_2(4) = z(4) = 0.2, & S_2(7) = z(7) = -0.1, \\ S_3(7) = z(7) = -0.1, & S_3(10) = z(10) = -0.2 \end{array} \quad (*.6)$$

Now, let us consider the second row of equations (*.6) in detail:

$$S_2(4) = z(4) = 0.2, \quad S_2(7) = z(7) = -0.1 \Rightarrow$$

$$\begin{cases} \frac{1}{6}M_2 \cdot \frac{(7-4)^3}{(7-4)} + \frac{1}{6}M_3 \cdot \frac{(4-4)^3}{(7-4)} + A_2(4-4) + B_2 = z_2 = 0.2 \; for \; x = 4 \\[2mm] \frac{1}{6}M_2 \cdot \frac{(7-7)^3}{(7-4)} + \frac{1}{6}M_3 \cdot \frac{(7-4)^3}{(7-4)} + A_2(7-4) + B_2 = z_3 = -0.1 \; for \; x = 7 \end{cases}$$

$$(*.7)$$

After some simplification we obtain

$$\begin{cases} \frac{1}{6}M_2 \cdot 3^2 + B_2 = z_2 = 0.2 \; for \; x = 4 \\[2mm] \frac{1}{6}M_3 \cdot 3^2 + A_2(7-4) + B_2 = z_3 = -0.1 \; for \; x = 7 \end{cases} \quad (*.7')$$

Finally:

$$\begin{cases} B_2 = z_2 - \dfrac{9}{6}M_2 = 0.2 - \dfrac{3}{2}M_2 \\ A_2 = \dfrac{1}{3}\left(z_3 - z_2 - \dfrac{3^2}{6}(M_3 - M_2)\right) = \dfrac{1}{3}(z_3 - z_2) - \dfrac{1}{2}(M_3 - M_2) = -\dfrac{3}{30} - \dfrac{1}{2}(M_3 - M_2) \end{cases}$$

$$(*.7'')$$

For the other equations in (*.6):

$$\begin{cases} B_1 = z_1 - \dfrac{9}{6}M_1 = 0.1 - \dfrac{9}{6}M_1 = 0.1 \\ A_1 = \dfrac{1}{3}\left(z_2 - z_1 - \dfrac{9}{6}(M_2 - M_1)\right) = \dfrac{1}{3}(z_2 - z_1) - \dfrac{1}{2}(M_2 - M_1) = \dfrac{1}{30} - \dfrac{1}{2}(M_2 - M_1) \end{cases}$$

$$(*.8)$$

hold and we further obtain

$$\begin{cases} B_3 = z_3 - \dfrac{9}{6}M_3 = -0.1 - \dfrac{3}{2}M_3 \\ A_3 = \dfrac{1}{3}\left(z_4 - z_3 - \dfrac{9}{6}(M_4 - M_3)\right) = \dfrac{1}{3}(z_4 - z_3) - \dfrac{1}{2}(M_4 - M_3) = -\dfrac{1}{30} + \dfrac{1}{2}M_3 \end{cases}$$

$$(*.8')$$

Thus, if we can determine the unknown second derivatives M_2, M_3, we have the complete spline. To ensure smooth passage between the functions $S_1(x), S_2(x)$, $S_3(x)$ it is necessary that the first derivatives of adjoining functions at measurement locations be equal:

$$\frac{d}{dx}S_1(4) = \frac{d}{dx}S_2(4), \quad \frac{d}{dx}S_2(7) = \frac{d}{dx}S_3(7)$$

$$(*.9)$$

This leads to

$$\begin{cases} \dfrac{1}{2}M_2 \cdot \dfrac{(4-1)^2}{(4-1)} + A_1 = -\dfrac{1}{2}M_2 \cdot \dfrac{(7-4)^2}{(7-4)} + \dfrac{1}{2}M_3 \cdot \dfrac{(4-4)^2}{(7-4)} + A_2, \quad for\ x=4, \\ -\dfrac{1}{2}M_2 \cdot \dfrac{(7-7)^2}{(7-4)} + \dfrac{1}{2}M_3 \cdot \dfrac{(7-4)^2}{(7-4)} + A_2 = -\dfrac{1}{2}M_3 \cdot \dfrac{(10-7)^2}{(10-7)} + A_3, \quad for\ x=7 \end{cases}$$

$$(*.9')$$

After simplification and setting A_1, A_2, A_3 as obtained above, we get

$$\begin{cases} \dfrac{3}{2}M_2 + A_1 = -\dfrac{3}{2}M_2 + A_2 \Rightarrow \dfrac{3}{2}M_2 + \dfrac{1}{3}\left(0.1 - \dfrac{9}{6}M_2\right) = -\dfrac{3}{2}M_2 + \dfrac{1}{3}\left(-0.3 - \dfrac{9}{6}(M_3 - M_2)\right) \\ \dfrac{3}{2}M_3 + A_2 = -\dfrac{3}{2}M_3 + A_3 \Rightarrow \dfrac{3}{2}M_3 + \dfrac{1}{3}\left(-0.3 - \dfrac{9}{6}(M_3 - M_2)\right) = -\dfrac{3}{2}M_3 + \dfrac{1}{3}\left(-0.1 + \dfrac{9}{6}M_3\right) \end{cases}.$$

Now we have two linear equations [or a linear system of equations (LSE)] with two unknown variables M_2, M_3. A well-known mathematical theorem from calculus ensures that a solution of this LSE exists. In our case the solution is given by $M_2 = 6/75$, $M_3 = 4/75$. Thus, after using these values in (*.7″), (*.8), and (*.8′) for evaluating the constants A_1, A_2, A_3 and B_1, B_2, B_3, the cubic spline (*.10) is calculated [see (*.1) and Fig. 3.3]:

$$S(x) = \begin{cases} \dfrac{-0.08}{6} \cdot \dfrac{(x-1)^3}{(4-1)} + 0.0733\,(x-1) + 0.1, & x \in [1,4] \,, \\[2mm] \dfrac{-0.08}{6} \cdot \dfrac{(7-x)^3}{(7-4)} + \dfrac{0.0533}{6} \cdot \dfrac{(x-4)^3}{(7-4)} - 0.1667\,(x-4) + 0.32, & x \in [4,7] \,, \\[2mm] \dfrac{0.0533}{6} \cdot \dfrac{(10-x)^3}{(10-7)} - 0.0067\,(x-7) - 0.18, & x \in [7,10] \end{cases} \qquad (*.10)$$

The common rule for calculating 1D-cubic splines reads as follows:

Let interval $[a,b]$ be divided into $N-1$ subintervals via $\Delta = \{a = x_1 < x_2 < \ldots < x_N = b\}$. The cubic spline is a twice-differentiable, piecewise cubic function of following form:

$$S(x) = \begin{cases} S_1(x), & x \in [x_1, x_2], \\ \ldots \\ S_{N-1}(x), & x \in [x_{N-1}, x_N] \end{cases} \qquad (3\text{-}6)$$

with

$$S(x) = M_j \frac{(x_{j+1} - x)^3}{6h_{j+1}} + M_{j+1} \frac{(x - x_j)^3}{6h_{j+1}} + A_j(x - x_j) + B_j, j = 1, \ldots, N-1$$

$$for \quad x \in [x_j, x_{j+1}] \quad with$$

$$h_{j+1} = x_{j+1} - x_j,$$

$$M_j = \frac{d^2}{dx^2} S(x_j), \quad M_1 = M_N = 0 \tag{3-7}$$

and $M_2, \ldots, M_{N-1}$ are solutions of the following LSE:

$$\begin{pmatrix} 2 & \lambda_2 & 0 & \ldots & & & \\ \mu_3 & 2 & \lambda_3 & 0 & \ldots & & \\ 0 & \mu_4 & 2 & \lambda_4 & 0 & \ldots & \\ \ldots & & & & & & \\ 0 & \ldots & 0 & \mu_{N-2} & 2 & \lambda_{N-2} \\ 0 & 0 & \ldots & 0 & \mu_{N-1} & 2 \end{pmatrix} \cdot \begin{pmatrix} M_2 \\ M_3 \\ \ldots \\ \ldots \\ \ldots \\ M_{N-1} \end{pmatrix} = \begin{pmatrix} d_2 \\ d_3 \\ \ldots \\ \ldots \\ \ldots \\ d_{N-1} \end{pmatrix} \tag{3-8}$$

where following abbreviations are used:

$$\lambda_j = \frac{h_{j+1}}{h_j + h_{j+1}}, \quad \mu_j = 1 - \lambda_j,$$

$$d_j = \frac{6}{h_j + h_{j+1}} \left\{ \frac{z_{j+1} - z_j}{h_{j+1}} - \frac{z_j - z_{j-1}}{h_j} \right\}, \quad j = 2, \ldots, N-1$$

$$A_j = \frac{z_{j+1} - z_j}{h_{j+1}} - \frac{h_{j+1}}{6}(M_{j+1} - M_j),$$

$$B_j = y_j - M_j \frac{h_{j+1}^2}{6},$$

$$j = 1, \ldots, N-1.$$

Remark: As noted above the solution of LSE (3-8) is ensured by theoretical considerations. Moreover, if the data come from an unknown function $f(x)$, the corresponding cubic spline $S(x)$ in a given interval $[a, b]$ converges to the function $f(x)$ in this interval. Owing this fact, the unwanted oscillations between data locations can be damped. These facts substantiate the obvious advantages of this approach. The main disadvantage of this method is the laborious numerical solution of (3-8) for large data sets. An additional measurement leads to a complete recalculation of the cubic spline.

For the two-dimensional case there are similar methods. The interested reader can find a detailed presentation of spline theory in Dierckx (1993). Some further interpolation and approximation methods are discussed in Chap. 4.

Problem 2A-ND *Regular data (measurements on a mesh/grid) should be interpolated (be described) with a functional relation roughly describing the data structure.*

Here the interpolation demand is omitted. Thus, a functional relation has to be obtained that describes the structure without exact reproduction of the z-values (measurements) at their locations. This approach is of use if measurements are distorted through random mistakes. This sort of interpolation helps to filter a so-called *external drift* in data. The key word for such approaches is "regression." More on this topic can be found in Sect. 3.2.1.

3.1.1.6 Polynomial Regression (1D-Case)

Often the simplest approach, *linear* regression, where a linear structure of the given data is assumed, is applied. We discuss the basic idea of this regression approach in detail in what follows and present the common rule for polynomial regressions later.

Example 3.1.1.1'''' (*1D-Case*) We return to the temporal measurements:

$$z_1 = z(1) = 0.1, \, z_2 = z(4) = 0.2, z_3 = z(7) = -0.1, \, z_4 = z(10) = -0.2$$

and look for a linear function $z(x) = ax + b$ that can describe the data structure and fits these data well. The parameters a and b have to be determined or estimated to obtain this linear function.

First, we must introduce a *goodness measure* for our approximation. There are an infinite number of lines going through and crossing the given points. Which of these is the best line? The answer comes from the great German mathematician Gauss, who proposed the following measure: the best line is the line that leads to *least squares* of differences obtained at the data locations. Exactly the squares and not any higher degrees of differences are considered because the linearity of the final system of equations should hold.

Let us formulate this problem in mathematical language. The parameters a and b have be determined so that the following function of two variables,

$$
\begin{aligned}
F(a,b) &= \sum_{i=1}^{4} (ax_i + b - z_i)^2 \\
&= (a \cdot 1 + b - 0.1)^2 + (a \cdot 4 + b - 0.2)^2 + (a \cdot 7 + b - (-0.1))^2 \qquad (*.1) \\
&\quad + (a \cdot 10 + b - (-0.2))^2 \\
&\to \min_{a,b}
\end{aligned}
$$

is minimized with respect to a and b.

From calculus we know that the following necessary condition must be fulfilled:

$$
\begin{cases}
\dfrac{\partial F}{\partial a} = \sum_{i=1}^{4} 2(ax_i + b - z_i)x_i = 0 \\
\dfrac{\partial F}{\partial b} = \sum_{i=1}^{4} 2(ax_i + b - z_i) \cdot 1 = 0
\end{cases}
\Rightarrow
\begin{cases}
\sum_{i=1}^{4} (ax_i + b - z_i)x_i = 0 \\
\sum_{i=1}^{4} (ax_i + b - z_i) \cdot 1 = 0
\end{cases}
\Rightarrow \qquad (*.2)
$$

$$
\begin{cases}
(a \cdot 1 + b - 0.1) \cdot 1 + (a \cdot 4 + b - 0.2) \cdot 4 + (a \cdot 7 + b - (-0.1)) \cdot 7 + (a \cdot 10 + b - (-0.2)) \cdot 10 = 0 \\
a \cdot 1 + b - 0.1 + a \cdot 4 + b - 0.2 + a \cdot 7 + b - (-0.1) + a \cdot 10 + b - (-0.2) = 0
\end{cases}
$$
$$(*.3)$$

It can be seen that this system of equations is linear with respect to the unknown parameters a and b. The following linear equation system is given:

$$
\begin{cases}
a \sum_{i=1}^{4} x_i^2 + b \sum_{i=1}^{4} x_i = \sum_{i=1}^{4} z_i x_i \\
a \sum_{i=1}^{4} x_i + b \cdot 4 = \sum_{i=1}^{4} z_i
\end{cases}
\Rightarrow
\begin{pmatrix} \sum_{i=1}^{4} x_i^2 & \sum_{i=1}^{4} x_i \\ \sum_{i=1}^{4} x_i & 4 \end{pmatrix} \cdot \begin{pmatrix} a \\ b \end{pmatrix} = \begin{pmatrix} \sum_{i=1}^{4} z_i x_i \\ \sum_{i=1}^{4} z_i \end{pmatrix}
$$

$$
\begin{cases}
a \cdot (1^2 + 4^2 + 7^2 + 10^2) + b \cdot (1 + 4 + 7 + 10) = 0.1 + 0.2 \cdot 4 + (-0.1) \cdot 7 + (-0.2) \cdot 10 \\
a \cdot (1 + 4 + 7 + 10) + b \cdot 4 = 0.1 + 0.2 - 0.1 - 0.2
\end{cases}
$$

The matrix representation leads to the solution:

$$\begin{pmatrix} \sum\limits_{i=1}^{4} x_i^2 & \sum\limits_{i=1}^{4} x_i \\ \sum\limits_{i=1}^{4} x_i & 4 \end{pmatrix} \cdot \begin{pmatrix} a \\ b \end{pmatrix} = \begin{pmatrix} \sum\limits_{i=1}^{4} z_i x_i \\ \sum\limits_{i=1}^{4} z_i \end{pmatrix} \Rightarrow \begin{pmatrix} a \\ b \end{pmatrix} = \begin{pmatrix} 166 & 22 \\ 22 & 4 \end{pmatrix}^{-1} \cdot \begin{pmatrix} -1.8 \\ 0.0 \end{pmatrix} = \begin{pmatrix} -0.04 \\ 0.22 \end{pmatrix}$$

$$(*.3')$$

Finally, we get $a = -0.04, b = 0.22$. If we assume a linear structure of the data, the following linear relation describes the relation between z and x:

$$z(x) = -0.04x + 0.22 \qquad (*.4)$$

Figure 3.4 shows (*.4) approximating the given data points. This linear function leads to the least squares of the differences between predicted and real z-values at the data locations. The sum of least squares of these differences can be evaluated using (*.1). With $a = -0.04$, $b = 0.22$ in (*.1), we get

$$Sum_{LR} = (-0.04 \cdot 1 + 0.22 - 0.1)^2 + (-0.04 \cdot 4 + 0.22 - 0.2)^2$$
$$+ (-0.04 \cdot 7 + 0.22 - (-0.1))^2 + (-0.04 \cdot 10 + 0.22 - (-0.2))^2 = 0.028$$
$$(*.5)$$

The value from (*.5) can be used as the goodness measure of our linear model for describing the data structure. Its comparison with

$$N \cdot (z_{max} - z_{min})^2 = 4 \cdot (0.2 - (-0.2))^2 = 0.64 \qquad (*.6)$$

is especially meaningful: If these values are almost equal, that is, if they have the same order in a mathematical sense, then our linear model is bad and should be

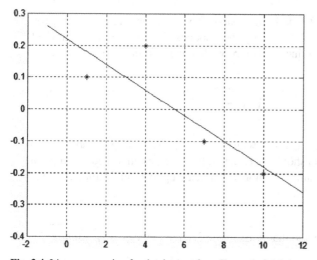

Fig. 3.4 Linear regression for the data set from Example 3.1.1.1

rejected. In our case we obtained an acceptable model because the value from (*.5) is smaller than the value from (*.6) by approximately one order of magnitude.

First, we formulate *the common rule for applying linear regression (1D-case):*

Given the data $z_1 = z(x_1), .., z_N = z(x_N)$, a line with the form $z(x) = ax + b$ has to be determined in such a way that the following necessary condition is fulfilled:

$$F(a,b) = \sum_{i=1}^{N} (ax_i + b - z_i)^2 \rightarrow \min_{a,b}$$

This condition leads to

$$\begin{cases} \dfrac{\partial F}{\partial a} = \sum_{i=1}^{N} 2(ax_i + b - y_i)x_i = 0 \\ \dfrac{\partial F}{\partial b} = \sum_{i=1}^{N} 2(ax_i + b - y_i) \cdot 1 = 0 \end{cases} \Rightarrow \begin{cases} \dfrac{\partial F}{\partial a} = \sum_{i=1}^{N} (ax_i + b - y_i)x_i = 0 \\ \dfrac{\partial F}{\partial b} = \sum_{i=1}^{N} (ax_i + b - y_i) \cdot 1 = 0 \end{cases}$$

and

$$\begin{cases} a\sum_{i=1}^{N} x_i^2 + b\sum_{i=0}^{N} x_i = \sum_{i=1}^{N} y_i x_i \\ a\sum_{i=1}^{N} x_i + b \cdot N = \sum_{i=1}^{N} y_i \end{cases} \Rightarrow \begin{pmatrix} \sum\limits_{i=1}^{N} x_i^2 & \sum\limits_{i=1}^{N} x_i \\ \sum\limits_{i=1}^{N} x_i & N \end{pmatrix} \cdot \begin{pmatrix} a \\ b \end{pmatrix} = \begin{pmatrix} \sum\limits_{i=1}^{N} z_i x_i \\ \sum\limits_{i=1}^{N} z_i \end{pmatrix}$$

The unknown parameters a and b are obtained by

$$\begin{pmatrix} a \\ b \end{pmatrix} = \begin{pmatrix} \sum\limits_{i=1}^{N} x_i^2 & \sum\limits_{i=1}^{N} x_i \\ \sum\limits_{i=1}^{N} x_i & N \end{pmatrix}^{-1} \cdot \begin{pmatrix} \sum\limits_{i=1}^{N} z_i x_i \\ \sum\limits_{i=1}^{N} z_i \end{pmatrix} \tag{3-9}$$

The line $z(x) = ax + b$ with these parameters is the one we are looking for. Moreover, we can calculate exactly the goodness measure (the accuracy) of our approximation by a linear relation. With a and b from (3-9) we define

$$Acc_{LR} = \sqrt{\frac{1}{N} \sum_{i=1}^{N} (ax_i + b - z_i)^2} \tag{3-9'}$$

If this accuracy is comparable with $z_{max} - z_{min}$, the assumption about a linear structure of the data should be rejected. In this case a modified assumption should be used. For example, a more complicated polynomial structure of the data can be assumed.

Remark It should be noted that in Example 3.1.1.1$''''$ $Acc_{LR} = \sqrt{\frac{1}{N} Sum_{LR}}$.

The common rule for using polynomial regression (1D-case) is formulated as follows:

Given the : $z_1 = z(x_1), .., z_N = z(x_N)$, a functional relation of the form $z(x) = a_M \cdot x^M + a_{M-1} \cdot x^{M-1} + \ldots + a_1 \cdot x + b$, $M \leq N-1$ has to be found that fulfills the following necessary condition:

$$F(a_1, \ldots, a_M, b) = \sum_{i=1}^{N} \left(a_M x_i^M + \ldots a_1 \cdot x_i + b - z_i\right)^2 \to \min_{a_1, \ldots, a_M, b}$$

This condition leads to:

$$\begin{cases} \dfrac{\partial F}{\partial a_M} = \sum_{i=1}^{N} 2\left(a_M x_i^M + \ldots + a_1 x_i + b - z_i\right) x_i^M = 0 \\ \ldots \\ \dfrac{\partial F}{\partial a_1} = \sum_{i=1}^{N} 2\left(a_M x_i^M + \ldots + a_1 x_i + b - z_i\right) x_i = 0 \\ \dfrac{\partial F}{\partial b} = \sum_{i=1}^{N} 2\left(a_M x_i^M + \ldots + a_1 x_i + b - z_i\right) \cdot 1 = 0 \end{cases} \Rightarrow$$

and, after some simplification, to

$$\begin{cases} a_M \sum_{i=1}^{N} x_i^{2M} + \ldots + a_1 \sum_{i=1}^{N} x_i^{M+1} + b \sum_{i=0}^{N} x_i^M = \sum_{i=1}^{N} z_i x_i^M \\ \ldots \\ a_M \sum_{i=1}^{N} x_i^{M+1} + \ldots + a_1 \sum_{i=1}^{N} x_i^2 + b \sum_{i=0}^{N} x_i = \sum_{i=1}^{N} z_i x_i \\ a_M \sum_{i=1}^{N} x_i^M + \ldots + a_1 \sum_{i=1}^{N} x_i + b \cdot N = \sum_{i=1}^{N} z_i \end{cases} \Rightarrow$$

and finally to the following solution of this LSE in matrix form:

$$\begin{pmatrix} a_M \\ \ldots \\ a_1 \\ b \end{pmatrix} = \begin{pmatrix} \sum_{i=1}^{N} x_i^{2M} & \ldots & \sum_{i=1}^{N} x_i^{M+1} & \sum_{i=1}^{N} x_i^M \\ \ldots & \ldots & \ldots & \ldots \\ \sum_{i=1}^{N} x_i^{M+1} & \ldots & \sum_{i=1}^{N} x_i^2 & \sum_{i=1}^{N} x_i \\ \sum_{i=1}^{N} x_i^M & \ldots & \sum_{i=1}^{N} x_i & N \end{pmatrix}^{-1} \cdot \begin{pmatrix} \sum_{i=1}^{N} z_i x_i^M \\ \ldots \\ \sum_{i=1}^{N} z_i x_i \\ \sum_{i=1}^{N} z_i \end{pmatrix} \qquad (3\text{-}10)$$

Thus, the unknown parameters $a_1, \ldots, a_M$ and b are obtained. With these parameters, the functional relation

$$z(x) = a_M \cdot x^M + a_{M-1} \cdot x^{M-1} + \ldots + a_1 \cdot x + b, \quad M \leq N-1$$

is identified.

The accuracy (goodness measure) of the polynomial approximation can be obtained from

$$ACC_{PR} = \sqrt{\frac{1}{N} \sum_{i=1}^{N} \left(a_M x_I^M + \ldots + a_1 x + b - z_i\right)^2} \qquad (3\text{-}10')$$

using the parameters $a_1, \ldots, a_M$ and b from (3-10') and a comparison with $z_{max} - z_{min}$.

Remark: For the present, (3-10) seems to be very simple, but the corresponding matrix can be irregular. Thus, some necessary conditions should be proved before trying to evaluate the inverse matrix. Moreover, numerical trouble can result from inverting large matrices.

For the case $N = M - 1$, a polynomial is obtained that fulfills the interpolation demand. It is identical with the one obtained by the Lagrange and the Newton interpolation methods. If the accuracy of (3-10') is comparable with $z_{max} - z_{min}$, the assumption about a polynomial structure of the data should be rejected. Analogously to the 1D-case, a more complicated polynomial structure of the data should be assumed.

3.1.1.7 Polynomial Regression (2D-Case)

Let us start with our familiar example and explain it in detail. Then we can present the common rule.

Example 3.1.1.2″ (2D-Case) Again, we consider the following measurements:

$$z_1 = z(x_1, y_1) = z(1, 0) = 0.1, \quad z_2 = z(x_2, y_2) = z(4, 0) = 0.2,$$
$$z_3 = z(x_3, y_3) = z(1, 1) = -0.1, \quad z_4 = z(x_4, y_4) = z(4, 1) = -0.2.$$

We assume that the data structure follows $z(x, y) = a_{10} \cdot x + a_{01} \cdot y + a_{00}$. A two-dimensional polynomial of this form has to be determined. Obviously, this polynomial is two dimensional of degree one. The basic idea of this approximation follows the approach for the 1D-case. We are looking for parameters a_{00}, a_{10}, a_{01} that lead to

$$F(a_{00}, a_{01}, a_{10}) = \sum_{i=1}^{4} \left(a_{10} x_i + a_{01} y_i + a_{00} - z_i\right)^2 \rightarrow \min_{a_{00}, a_{10}, a_{01}} \qquad (*.1)$$

Using partial derivatives with respect to these parameters, we get:

$$\begin{cases} \dfrac{\partial F}{\partial a_{10}} = \sum_{i=1}^{4} 2\left(a_{10}x_i + a_{01}y_i + a_{00} - z_i\right)x_i = 0 \\[2mm] \dfrac{\partial F}{\partial a_{01}} = \sum_{i=1}^{4} 2\left(a_{10}x_i + a_{01}y_i + a_{00} - z_i\right)y_i = 0 \\[2mm] \dfrac{\partial F}{\partial a_{00}} = \sum_{i=1}^{4} 2\left(a_{10}x_i + a_{01}y_i + a_{00} - z_i\right)\cdot 1 = 0 \end{cases} \Rightarrow \begin{pmatrix} a_{10} \\ a_{01} \\ a_{00} \end{pmatrix} = \begin{pmatrix} \sum_{i=1}^{4} x_i^2 & \sum_{i=1}^{4} y_i x_i & \sum_{i=1}^{4} x_i \\ \sum_{i=1}^{4} y_i x_i & \sum_{i=1}^{4} y_i^2 & \sum_{i=1}^{4} y_i \\ \sum_{i=1}^{4} x_i & \sum_{i=1}^{4} y_i & 4 \end{pmatrix}^{-1} \cdot \begin{pmatrix} \sum_{i=1}^{4} x_i z_i \\ \sum_{i=1}^{4} y_i z_i \\ \sum_{i=1}^{4} z_i \end{pmatrix}$$

With the given values from our data set, we obtain:

$$\begin{pmatrix} a_{10} \\ a_{01} \\ a_{00} \end{pmatrix} = \begin{pmatrix} 34 & 5 & 10 \\ 5 & 2 & 2 \\ 10 & 2 & 4 \end{pmatrix}^{-1} \cdot \begin{pmatrix} 0 \\ -0.3 \\ 0 \end{pmatrix} = \begin{pmatrix} 0 \\ -0.3 \\ 0.15 \end{pmatrix} \qquad (*.2)$$

and the functional relation is identified as a plane or a two-dimensional polynomial of degree one of the form $z(x, y) = 0 \cdot x - 0.3 \cdot y + 0.15$ (see Fig. 3.5).

The common rule for two-dimensional polynomial regression reads as follows:

Let the following data be given: $z_1 = z(x_1, y_1), \ldots, z_N = z(x_N, y_N)$. A functional relation of the form

$$z(x, y) = \sum_{k=0}^{K} \sum_{l=0}^{L} a_{kl} \cdot x^k \cdot y^l, \quad (K+1)(L+1) \leq N$$

is searched for in such a way that the following necessary condition is fulfilled:

$$F(a_{kl} : k = 0 \ldots K, l = 0 \ldots L) = \sum_{i=1}^{N} \left(\sum_{k=0}^{K} \sum_{l=0}^{L} a_{kl} \cdot x_i^k \cdot y_i^l - z_i \right)^2 \to \min_{a_{kl}:k=0\ldots K,\, l=0\ldots L}$$

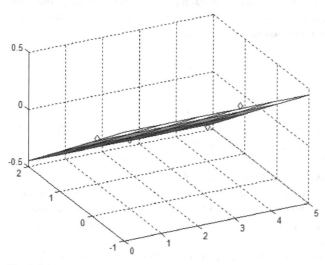

Fig. 3.5 Two-dimensional regression by a plane for the data from Example 3.1.1.2

This condition leads to

$$
\begin{cases}
\dfrac{\partial F}{\partial a_{k^*l^*}} = \sum_{i=1}^{N} 2 \left(\sum_{k=0}^{K} \sum_{l=0}^{L} a_{kl} x_i^k y_i^l - z_i \right) x_i^{k^*} y_i^{l^*} = 0 \\
k^* = 0 \dots K,\, l^* = 0 \dots L
\end{cases}
$$

and after some simplification of the $(K+1)(L+1)$ equations to the following LSE:

$$
\begin{cases}
\sum_{k=0}^{K} \sum_{l=0}^{L} a_{kl} \sum_{i=1}^{N} x_i^{k+k^*} y_i^{l+l^*} = \sum_{i=1}^{N} z_i x_i^{k^*} y_i^{l^*} \\
k^* = 0 \dots K,\, l^* = 0 \dots L
\end{cases}
\tag{3-11}
$$

The solution of this LSE leads to parameters a_{kl}, $k = 0 \dots K$, $l = 0 \dots L$.

The accuracy of the polynomial approximation the data can be described by

$$
Acc_{PR} = \sqrt{\frac{1}{N} \sum_{i=1}^{N} \left(\sum_{k=0}^{K} \sum_{l=0}^{L} a_{kl} x_i^k y_i^l - z_i \right)^2}
\tag{3-11'}
$$

with parameters that solve (3-11). For model validation this value should be compared with $z_{max} - z_{min}$.

Remark: For the case $N = (M+1)(L+1)$, a polynomial is obtained that fulfills the interpolation demand. The matrix of the LSE (3-11) can be irregular. Thus, some necessary conditions should be proved before trying to solve the LSE. Even in the case of a regular matrix there can be numerical trouble, for example, for a large N.

Following the 1D-case, the assumption about a polynomial structure of the data should be rejected if the accuracy is comparable with $z_{max} - z_{min}$. More sophisticated functional structures are presented in Sect. 3.2.1 and Chap. 4.

3.1.1.8 B-Splines (1D-Case)

The designation "B-Splines" goes back to the French mathematician Bezier. The basic idea of the approach comes from an *algorithm of Casteljau* based on repeated linear interpolations. This method is used successfully in modern CAD-tools. B-splines are well suited for applications that should be performed without considering coordinates. B-splines and B-surfaces are parametrically determined.

So-called B-curves are determined using one parameter, t. More general B-surfaces correspond to interpolation with two parameters, u and v. Changing the coordinate system (scaling the axes and further affine transformations) do not influence the basic algorithm.

Sometimes this approach is called the *end-point-interpolation method*. An explanation is given below. Now, we start constructing *B-curves* of degree N in the three-dimensional space E^3. For plane curves in E^2 the coordinate y should be omitted in the following equations.

We formulate *the common rule for constructing B-curves* as follows:

Let $b_0 = (x_0, y_0, z_0)^T$, $b_1 = (x_1, y_1, z_1)^T$, ..., $b_N = (x_N, y_N, z_N)^T \in E^3$ be $N+1$ data points and $t \in (-\infty, \infty)$ be a parameter. After linearly interpolating for N-times with a fixed parameter t with

$$b_i^r(t) = (1-t)b_i^{r-1} + tb_{i+1}^{r-1},$$
$$r = 1, \ldots, N$$
$$i = 0, \ldots, N-r \tag{3-12}$$
$$b_i^0 = b_i.$$

the point $b_0^N(t)$ finally obtained is a point on the B-curve $b^N(t)$ of degree N.

Remark: Remember that point numbering begins with zero. Thus, we have $N+1$ data points. Relation (3-12) refers to all coordinates of the data, namely, x, y, z (see Example 3.1.1.4).

The B-curve $b^N(t)$ is parametric and corresponds to the point b_0 for $t = 0$ and to the point b_N for $t = 1$. This curve winds through the other points (see Fig. 3.6), which explains the fact that interpolation with B-splines is called end-point interpolation. The polygon based on the points $b_0, b_1, \ldots, b_N \in E^3$ is called the *Bezier polygon* or the *control polygon* of the B-curve $b^N(t)$.

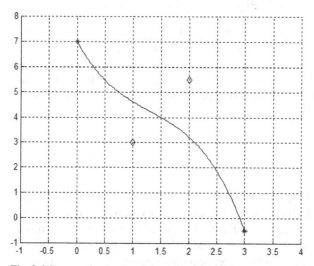

Fig. 3.6 B-curve for the data from example 3.1.1.4

Example 3.1.1.4 (1D-Case) The following temporal measurements are given:

$$z_1(0) = 7, \ z_2(1) = 3, \ z_3(2) = 5.5, \ z_4(3) = -0.5.$$

With the terms from (3-12) this corresponds to

$$b_0 = \begin{pmatrix} 0 \\ 7 \end{pmatrix}, \ b_1 = \begin{pmatrix} 1 \\ 3 \end{pmatrix}, \ b_2 = \begin{pmatrix} 2 \\ 5.5 \end{pmatrix}, \ b_3 = \begin{pmatrix} 3 \\ -0.5 \end{pmatrix}.$$

Now, how does the parametric curve describing this structure with fixed endpoints actually look? Here $N = 3$. Thus, we need three interpolation steps for obtaining the final B-curve from (3-12). The first step leads to

$$b_0^1 = (1-t)b_0 + tb_1 = (1-t)\begin{pmatrix} 0 \\ 7 \end{pmatrix} + t\begin{pmatrix} 1 \\ 3 \end{pmatrix} = \begin{pmatrix} t \\ 7-4t \end{pmatrix},$$

$$b_1^1 = (1-t)b_1 + tb_2 = (1-t)\begin{pmatrix} 1 \\ 3 \end{pmatrix} + t\begin{pmatrix} 2 \\ 5.5 \end{pmatrix} = \begin{pmatrix} 1+t \\ 3+2.5t \end{pmatrix}, \qquad (*.1)$$

$$b_2^1 = (1-t)b_2 + tb_3 = (1-t)\begin{pmatrix} 2 \\ 5.5 \end{pmatrix} + t\begin{pmatrix} 3 \\ -0.5 \end{pmatrix} = \begin{pmatrix} 1+t \\ 5.5-6t \end{pmatrix}.$$

These are three complementary points, which belong to the lines $[b_0, b_1]$, $[b_1, b_2]$, $[b_2, b_3]$ and divide them into identical proportions depending on t. In the second step we get

$$b_0^2 = (1-t)b_0^1 + tb_1^1 = (1-t)\begin{pmatrix} t \\ 7-4t \end{pmatrix} + t\begin{pmatrix} 1+t \\ 3+2.5t \end{pmatrix} = \begin{pmatrix} 2t \\ 6.5t^2 - 8t + 7 \end{pmatrix},$$

$$b_1^2 = (1-t)b_1^1 + tb_2^1 = (1-t)\begin{pmatrix} 1+t \\ 3+2.5t \end{pmatrix} + t\begin{pmatrix} 1+t \\ 5.5-6t \end{pmatrix} = \begin{pmatrix} 1+t \\ -8.5t^2 + 5t + 3 \end{pmatrix}.$$

$$(*.2)$$

These two points are placed at the lines $[b_0^1, b_1^1]$, $[b_1^1, b_2^1]$ and divide them into equal proportions depending on t. After the last step, the parametric B-curve is identified as

$$b_0^3 = (1-t)b_0^2 + tb_1^2 = (1-t)\begin{pmatrix} 2t \\ 6.5t^2 - 8t + 7 \end{pmatrix} + t\begin{pmatrix} 1+t \\ -8.5t^2 + 5t + 3 \end{pmatrix} \qquad (*.3)$$

and is as shown in Fig. 3.6.

Simple implementation and independence of coordinate systems are advantages of this approach. Like a designer, one can work with spatial terms "left," "right," "top," and "end" of a curve without a fixed relationship to coordinates. A rough approach can be improved to a smoothed, snaking line by using B-splines. Moreover, some approaches—polygons with identical end-points and the same number of points—should be compared based on their analytical parametric form given by

(3-12). For example, a mean polygon can be constructed in this way. One minor disadvantage might be the unconventional parametric representation of B-splines.

3.1.1.9 B-Splines (2D-Case)

Here the basic idea corresponds to an iterated bilinear interpolation. We begin with some definitions. Let b_{00}, b_{01}, b_{10}, b_{11} be four points in E^3. A hyperbolic paraboloid allows the following parametric representation:

$$z(u,v) = \sum_{i=0}^{1} \sum_{j=0}^{1} b_{ij} B_i^1(u) B_j^1(v), \quad mit \quad B_k^1(w) = (1-w)^{1-k} w^k,$$
$$k = 0, 1; \quad w = u, v$$

In matrix form this corresponds to

$$z(u,v) = \begin{bmatrix} 1-u & u \end{bmatrix} \begin{bmatrix} b_{00} & b_{01} \\ b_{10} & b_{11} \end{bmatrix} \begin{bmatrix} 1-v \\ v \end{bmatrix}$$

Now *the common rule for constructing B-curves* reads as follows:

Let $\{b_{ij}\}_{i,j=0}^{N} = \left\{ (x_{ij}, y_{ij}, z_{ij})^T \right\}_{i,j=0}^{N}$ be $(N+1)(N+1)$ three-dimensional data points (measurements on a square grid) and $(u,v) \in R^2$ be two parameters. After iterating linear interpolations for N-times with fixed parameters $(u,v) \in R^2$ with

$$b_{ij}^{rr}(u,v) = \begin{bmatrix} 1-u & u \end{bmatrix} \begin{bmatrix} b_{ij}^{r-1r-1} & b_{ij+1}^{r-1r-1} \\ b_{i+1j}^{r-1r-1} & b_{i+1j+1}^{r-1r-1} \end{bmatrix} \begin{bmatrix} 1-v \\ v \end{bmatrix} \quad (3\text{-}13)$$
$$r = 1, \ldots, N; \quad i, j = 0, \ldots, N-r$$

with

$$b_{ij}^{00} = b_{ij}$$

the point $b_{00}^{NN}(u,v)$ finally obtained is a point on the B-surface $b^{NN}(u,v)$.

Remark: Point numbering starts with zero, so we have $(N+1)(N+1)$ data points in all. Equation (3-13) refers to all the coordinates of the data, namely, x, y, z (see Example 3.1.1.5).

The B-surface $b^{NN}(t)$ is parametric and corresponds to the point b_{00} for $(u, v) = (0, 0)$, to b_{10} for $(u, v) = (1, 0)$, to point b_{10} for $(u, v) = (0, 1)$, and to the point b_{11} for $(u, v) = (1, 1)$. This surface also winds through the other points (see Fig. 3.7), which, analogously to the 1D-case, indicates end-point interpolation. The polygon based on points $\{b_{ij}\}_{i,j=0}^N$ is called the *Bezier polygon* or the *control polygon* of the B-surface $b^{NN}(u, v)$.

There is a generalization for nonsquare grids for which further details can be found in Farin (1993).

Example 3.1.1.5 (2D-Case) We consider the following nine spatial measurements on a 3×3 mesh:

$$b_{00} = \begin{pmatrix} 0 \\ 0 \\ 1 \end{pmatrix}, \; b_{01} = \begin{pmatrix} 0 \\ 1 \\ -2 \end{pmatrix}, \; b_{02} = \begin{pmatrix} 0 \\ 2 \\ 5 \end{pmatrix}, \; b_{10} = \begin{pmatrix} 1 \\ 0 \\ 3 \end{pmatrix}, \; b_{11} = \begin{pmatrix} 1 \\ 1 \\ 4 \end{pmatrix}, \; b_{12} = \begin{pmatrix} 1 \\ 2 \\ 1 \end{pmatrix},$$

$$b_{20} = \begin{pmatrix} 2 \\ 0 \\ -1 \end{pmatrix}, \; b_{21} = \begin{pmatrix} 2 \\ 1 \\ -2 \end{pmatrix}, \; b_{22} = \begin{pmatrix} 2 \\ 2 \\ 2 \end{pmatrix}$$

How can we identify a parametric B-surface that fits this data set? We use (3-13) for the first step of a bilinear interpolation:

$$b_{00}^{11}(u, v) = \begin{bmatrix} 1 - u & u \end{bmatrix} \begin{bmatrix} b_{00} & b_{01} \\ b_{10} & b_{11} \end{bmatrix} \begin{bmatrix} 1 - v \\ v \end{bmatrix},$$

$$b_{01}^{11}(u, v) = \begin{bmatrix} 1 - u & u \end{bmatrix} \begin{bmatrix} b_{01} & b_{02} \\ b_{11} & b_{12} \end{bmatrix} \begin{bmatrix} 1 - v \\ v \end{bmatrix},$$

$$(*.1)$$

$$b_{10}^{11}(u, v) = \begin{bmatrix} 1 - u & u \end{bmatrix} \begin{bmatrix} b_{10} & b_{12} \\ b_{20} & b_{12} \end{bmatrix} \begin{bmatrix} 1 - v \\ v \end{bmatrix},$$

$$b_{11}^{11}(u, v) = \begin{bmatrix} 1 - u & u \end{bmatrix} \begin{bmatrix} b_{11} & b_{12} \\ b_{21} & b_{22} \end{bmatrix} \begin{bmatrix} 1 - v \\ v \end{bmatrix},$$

The evaluations in (*.1) should be performed separately for each coordinate x, y, z. For example, for the first equation in (*.1) we obtain:

$$x_{00}^{11}(u, v) = \begin{bmatrix} 1 - u & u \end{bmatrix} \begin{bmatrix} 0 & 0 \\ 1 & 1 \end{bmatrix} \begin{bmatrix} 1 - v \\ v \end{bmatrix} = \begin{bmatrix} 1 - u & u \end{bmatrix} \begin{bmatrix} 0 \\ 1 \end{bmatrix} = u,$$

$$y_{00}^{11}(u, v) = \begin{bmatrix} 1 - u & u \end{bmatrix} \begin{bmatrix} 0 & 1 \\ 0 & 1 \end{bmatrix} \begin{bmatrix} 1 - v \\ v \end{bmatrix} = \begin{bmatrix} 1 - u & u \end{bmatrix} \begin{bmatrix} v \\ v \end{bmatrix} = v,$$

$$z_{00}^{11}(u, v) = \begin{bmatrix} 1 - u & u \end{bmatrix} \begin{bmatrix} 1 & -2 \\ 3 & 4 \end{bmatrix} \begin{bmatrix} 1 - v \\ v \end{bmatrix} = \begin{bmatrix} 1 - u & u \end{bmatrix} \begin{bmatrix} 1 - 3v \\ 3 + v \end{bmatrix} = 1 - 3v + 2u + 4uv$$

In the second step we interpolate the results of (*.1):

$$b_{00}^{22}(u, v) = \begin{bmatrix} 1 - u & u \end{bmatrix} \begin{bmatrix} b_{00}^{11} & b_{01}^{11} \\ b_{10}^{11} & b_{11}^{11} \end{bmatrix} \begin{bmatrix} 1 - v \\ v \end{bmatrix} \qquad (*.2)$$

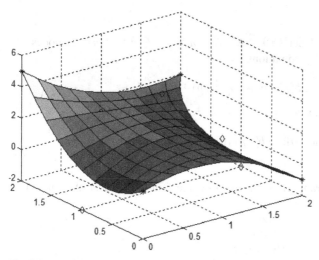

Fig. 3.7 B-surface for the data set from Example 3.1.1.5. The end-points are marked by stars

The parametric surface in (*.2) is the B-surface we looked for, shown in Fig. 3.7.

Problem 2.B *Irregularly spaced data (chaotically distributed measurements) should be interpolated (or described) with a functional relation.*

Regression approaches as presented for Problem 2.A remain relevant for the 1D- as well as the 2D-case. More details about regression approaches can be found in Sect. 3.2.1 and in Chap. 4. The construction of B-curves corresponds to that for regular data and was described earlier.

Determining B-surfaces requires an additional step. We start by triangulating the sample points, a procedure that was explained for Problem 2.B.. After covering the definition area of the measurements completely, B-splining can be performed. Here special so-called *triangular B-splines* are used.

The reason for applying special weights—or parameters here—$p = (u, v, w)$, $u + v + w = 1$ corresponding to an inner coordinate system over a triangle is analogous to the approach for bilinear interpolation over a triangular mesh, as demonstrated in 3.1.1.3. For instance, this yields

$$u = \frac{A^1}{A^1 + A^2 + A^3}, \quad v = \frac{A^2}{A^1 + A^2 + A^3}, \quad w = \frac{A^3}{A^1 + A^2 + A^3}$$

for point P_0 in Fig. 3.2. On other hand, P_0 takes the coordinates (u, v, w), which relate to an inner coordinate system over triangle $P_1 P_2 P_3$.

Now, let us formulate *the common rule for constructing triangular B-surfaces*: Following Farin (1993), we use the abbreviations $e_1 = (1, 0, 0)$, $e_2 = (0, 1, 0)$, $e_3 = (0, 0, 1)$, $b_{ijk} = b_i$, $|i| = i + j + k$, $i, j, k > 0$.

Let there be a control net $\{b_i\} \in E^3$, $|i| = n$ with $^1/_2(n+1)(n+2)$ nodes. Using iterated linear interpolations

$$b_i^r(p) = ub_{i+e_1}^{r-1}(p) + vb_{i+e_2}^{r-1}(p) + wb_{i+e_3}^{r-1}(p),$$

$$p = (u, v, w), \quad r = 1 \ldots n, \quad |i| = n - r, \quad b_i^0(p) = b_i$$

(3-14)

we obtain a parametric surface $b_0^n(p)$ (triangular B-spline) of degree n.

A control net for parametric surfaces of degree four is based on the following nodes:

$$b_{040}$$
$$b_{031}b_{130}$$
$$b_{022}b_{121}b_{220}$$
$$b_{013}b_{112}b_{211}b_{310}$$
$$b_{004}b_{103}b_{202}b_{301}b_{400}$$

For more information about B-splines, see Farin (1993).

3.1.2 Stochastic Point of View: Methods of Geostatistics

Opinions about "randomness" in the world of science vary from firm belief to absolute rejection. However, methods based on stochastics or statistics might be suitable for such real problems, where a mathematical model taking randomness into account seems to be logical.

Complicated geometrical structures come from many areas of science and technology. For example, geological structures, biological tissues, sections of porous media, and oil fields often require statistical analyses. There are different mathematical areas that provide such models and methods. One of these is called geostatistics.

In the 1950s geostatistics became a rapidly evolving branch of applied mathematics and statistics, starting in the mining industry, as its methods were originally developed to improve the calculations of ore reserves. "Kriging," which is the generic term for various interpolation methods in geostatistics, is named for D. G. Krige, a mining engineer who, together with the statistician H. S. Sichel, was among the pioneers in this research area. Both of them worked in South Africa in the early 1950s. Later in that decade the center of geostatistical research moved to France, to the Commissariat de l'Energie Anatomique. Geostatistics earned the status of a scientific discipline owing to the work of G. Matheron, who refined Krige's methods, introduced the so-called *regionalized variable*, and developed the major concepts of the theory for estimating resources.

In the 1970s these geostatistical concepts became well known in other branches of the earth sciences. They are popular in industry and as well as in economics because there are problems that need evaluations of spatially and temporally—

sometimes spatio-temporally—distributed and even correlated data. However, the methods of linear geostatistics were actually not absolutely novel in mathematics. The estimation theory in statistics involves familiar methods. "Collocation" from geodesy and classical regression analysis are also popular. These methods are based on the *least-squares method* developed by the great mathematician C. F. Gauss, and they were later adapted to different research areas for a wide variety of problem formulations encountered in practice. However, the problem of choices—albeit one that is more philosophical and less mathematical—remains: the choice of the workspace, of the primary assumptions, and of viewpoint.

Currently, geostatistics embraces various models and methods. Moreover, owing to recent developments from mathematical statistics and numerical analysis as well as better and faster computers, their numbers continue to increase. In this book we offer a short overview of the basic concepts and present typical applications of some kriging methods. For more details concerning geostatistics we recommend the book by H. Wackernagel (1995).

We begin with some important definitions in geostatistics. We assume that the reader possesses some fundamental knowledge of stochastics, found, for example, in K.L. Chung (1968).

Definition 3.1.2-1 $\{Z(x)\}$, $x \in R^n$ is called a *random field* if $Z(x)$ is a *random variable* for each $x \in R^n$. If $x = t \in R \Rightarrow Z(t)$ can be called a *random process*.

Definition 3.1.2-2 A single realization $\{z(x)\}$, $x \in R^n$ of a random field, $\{Z(x)\}$, $x \in R^n$, is called a *regionalized variable*.

The difference between the terms "random field" and "regionalized variable" can be easily explained. Throw a coin three times. If its sides are marked with 1 and 0, then there are the following eight possible realizations of this random process (for a discrete time scale corresponding to each throw): $[0, 0, 0]$, $[0, 0, 1]$, $[0, 1, 0]$, $[1, 0, 0]$, $[0, 1, 1]$, $[1, 0, 1]$, $[1, 1, 0]$, and $[1, 1, 1]$. But we obtain only one of these realizations. A regionalized variable from this process is just one of these realizations.

Definition 3.1.2-3 The following equation is called the *covariance function* of the random field $\{Z(x)\}$, $x \in R^n$:

$$C(x, h) = Cov(Z(x), Z(h)) = E\{(Z(x) - EZ(x))(Z(x+h) - EZ(x+h))\}.$$

Remark: The covariance function is a vector function that depends on many variables. The *mean* of the corresponding random variable, that is, the product of random differences, is denoted by E in Definition 3.1.2-3. For simplification the following two assumptions should be made:

1. First-order stationarity: $EZ(x) = EZ(x+h) = \mu = const$. This means that the mean of the random field is constant and that the mean value is the same at any point in the field.

2. Second-order stationarity: $C(x, h) = C(|h|)$. This means that the covariance between any pair of locations depends on the length of the distance vector h that separates them. In this case the covariance function is a function of a single variable that depends only on the distance; therefore, second-order stationarity eliminates directional effects.

From a practical point of view the covariance function affects estimation in the same way as the influence function discussed in Sect. 3.1.1. But there are more complicated conditions for its construction [see Yaglom (1986) for further details].

Definition 3.1.2-4 A normalized covariance function is called a *correlation function*:

$$\rho(h) = \frac{C(h)}{C(o)}.$$

Definition 3.1.2-5 For random variables $Z_1 = Z(x_1), \dots, Z_N = Z(x_N)$ the following matrix is called a *variance-covariance matrix*:

$$\begin{bmatrix} c_{11} & c_{12} & \cdots & c_{1N} \\ c_{21} & c_{22} & \cdots & c_{2N} \\ \cdots & & & \cdots \\ c_{N1} & c_{N2} & \cdots & c_{NN} \end{bmatrix}$$

with

$$c_{ii} = C(0) = Var(Z_i); \quad c_{ij} = Cov(Z_i, Z_j) = EZ_i Z_j - EZ_i EZ_j, \quad i, j = 1, \dots, N.$$

The basic concept of stochastic interpolation broadly corresponds to that of deterministic interpolation using an influence function. Therefore, we have to fit a weighted average considering information from location points of the regionalized variable to obtain an estimation of a certain value at a fixed point (x_0, y_0) or, in short, x_0.

This basic idea can be explained using a simple example. Let $[z(x_1, y_1), z(x_2, y_2)]$ be a regionalized variable that can, for example, be the result of two drillings from an oil field. Or think about measurements of, for instance, temperature or soil parameters. However, we are looking for weights, w_i, $i = 1, 2$, to predict at the point (x_0, y_0). Like the idea of the generalized mean from Sect. 3.1.1, this means that

$$\hat{Z}(x_0, y_0) = \frac{w_1 \cdot z(x_1, y_1) + w_2 \cdot z(x_2, y_2)}{w_1 + w_2} = \alpha_1 z(x_1, y_1) + \alpha_2 z(x_2, y_2), \alpha_i = \frac{w_i}{w_1 + w_2}, i = 1, 2$$

These weights, α_i, $i = 1, 2$, should depend on the distance to the point of prediction: closer points should have a stronger influence on the prediction value. The estimator is marked with a hat, $\hat{Z}(x_0, y_0)$, in order to distinguish it from the true but unknown value of random variable $Z(x_0, y_0)$. As a measure of "closeness" of this estimator to the true value we can use

$$Var\left(Z(x_0, y_0) - \hat{Z}(x_0, y_0)\right) \to \min_{\alpha_1, \alpha_2}$$

The variance of the random vector in brackets can be understood here in some sense as its "stochastic length." With the assumption of first-order stationarity, we can assume that the mean of this vector is equal to zero. But of course there are many different random vectors that fulfill this assumption. Minimizing the variance, we restrict the class of these vectors and hope to obtain a meaningful estimation. The estimation procedure must be based on knowledge of the covariances among the random variables at the different points.

3.1.2.1 Simple Kriging

We start with an approach called *kriging with a known mean*. Let $\{Z(x)\}$, $x \in R^n$ be a stationary (first and second order) random field. The designation x refers to all corresponding coordinates and is used for simplification. We repeat that stationarity means that

$$\begin{aligned} EZ(x+h) &= EZ(x) = \mu, \\ Cov(Z(x+h), Z(x)) &= C(|h|). \end{aligned} \tag{3-15}$$

where the mean of the random field is known and constant. A *covariance model* should be chosen, and this is discussed in detail in a remark below. Here we assume that a model for the covariance function is given. We denote the regionalized variable by $[z(x_1), \ldots, z(x_N)]$. For the estimator at the point x_0 we choose the following form:

$$\hat{Z}(x_0) = \mu + \sum_{k=1}^{N} \alpha_k (Z(x_k) - \mu) \tag{3-16}$$

Obviously, first-order stationarity is fulfilled for this estimator:

$$E\hat{Z}(x_0) = \mu + E\left(\sum_{k=1}^{N} \alpha_k (Z(x_k) - \mu)\right) = \mu + \sum_{k=1}^{N} \alpha_k E(Z(x_k) - \mu)$$

$$= \mu + \sum_{k=1}^{N} \alpha_k \left(\underbrace{EZ(x_k)}_{\mu} - \mu\right) = \mu \tag{3-17}$$

Now, the variance has to be minimized:

$$Var\left(Z(x_0) - \hat{Z}(x_0)\right) \to \min \tag{3-18}$$

$$Var\left(Z\left(x_0\right)-\hat{Z}\left(x_0\right)\right)=E\left(Z\left(x_0\right)-\hat{Z}\left(x_0\right)\right)^2-\underbrace{E^2\left(Z\left(x_0\right)-\hat{Z}\left(x_0\right)\right)}_{0}$$

$$=E\left(Z\left(x_0\right)-\mu-\sum_{k=1}^{N}\alpha_k\left(Z\left(x_k\right)-\mu\right)\right)^2=E\left(Z\left(x_0\right)-\mu\right)^2$$

$$+\sum_{i=1}^{N}\sum_{j=1}^{N}\alpha_i\alpha_jE\left\{\left(Z\left(x_i\right)-\mu\right)\left(Z\left(x_j\right)-\mu\right)\right\}$$

$$-2\sum_{i=1}^{N}\alpha_iE\left\{\left(Z\left(x_0\right)-\mu\right)\left(Z\left(x_i-\mu\right)\right)\right\}$$

$$Var\left(Z\left(x_0\right)-\hat{Z}\left(x_0\right)\right)=c\left(0\right)+\sum_{i=1}^{N}\alpha_i\sum_{j=1}^{N}\alpha_jc_{ij}-2\sum_{i=1}^{N}\alpha_ic_{i0} \tag{3-18'}$$

with parameters calculated using the covariance model

$$c\left(0\right)=C\left(|0|\right),\quad c_{ij}=Cov\left(Z\left(x_i\right),Z\left(x_j\right)\right),\quad c_{i0}=Cov\left(Z\left(x_i\right),Z\left(x_0\right)\right),\quad i,j=1,\ldots,N.$$

The variance in (3-18') is a function of N variables, that is, the interesting weights. After partial differentiation with respect to these weights we get

$$Var\left(Z\left(x_0\right)-\hat{Z}\left(x_0\right)\right)\rightarrow\min_{\alpha_1,\ldots,\alpha_N}\Rightarrow$$

$$\frac{d}{d\alpha_k}\left\{\sum_{i=1}^{N}\alpha_i\sum_{j=1}^{N}\alpha_jc_{ij}-2\sum_{i=1}^{N}\alpha_ic_{i0}\right\}=0,\,k=1\ldots N$$

and finally

$$\sum_{j=1}^{N}\alpha_jc_{kj}=c_{k0},\quad k=1,\ldots,N \tag{3-19}$$

Equations (3-19) form a LSE that in matrix form corresponds to

$$\begin{bmatrix} c_{11} & c_{12} & \cdots & c_{1N} \\ c_{21} & c_{22} & \cdots & c_{2N} \\ \cdots & & \cdots & \\ c_{N1} & c_{N2} & \cdots & c_{NN} \end{bmatrix}\begin{bmatrix} \alpha_1 \\ \alpha_2 \\ \cdots \\ \alpha_N \end{bmatrix}=\begin{bmatrix} c_{10} \\ c_{20} \\ \cdots \\ c_{N0} \end{bmatrix}. \tag{3-19'}$$

Obviously, solving the LSE (3-19') leads directly to the weights that we are looking for. If we denote these weights by $\left[\alpha_1^L,\ldots\alpha_N^L\right]^T$, then the estimation is given by

$$\hat{z}\left(x_0\right)=\mu+\sum_{k=1}^{N}\alpha_k^L\left(z\left(x_k\right)-\mu\right) \tag{3-16'}$$

In addition to the predicted z-value, we also obtain the so-called *kriging variance*, an accuracy measure highly dependent on the chosen model. Considering (3-18′) with the weights from (3-19′) leads to

$$\sigma_{SK}^2 = c(0) - \sum_{k=1}^{N} \alpha_k^L c_{k0} \tag{3-20}$$

If the measurement locations are scattered irregularly in the space it is worthwhile to produce a map of the kriging variance as a complement to the map of the estimated or kriged estimates. This gives us an appreciation of the varying precision of the kriged estimates that are due to the irregular locations of the values of the regionalized variable.

Now let us test how simple kriging works by an example.

Example 3.1.2.1 We use the following values of a regionalized variable $z(0, 0.2) = 1$, $z(1.2, -0.9) = -3$ and also assume that $E(Z) = \mu = 2$ is known. We determine a model and then look for an estimation at the point $(x_0, y_0) = (0, 0)$ in the form

$$\hat{Z} = \mu + \alpha_1 (Z_1 - \mu) + \alpha_2 (Z_2 - \mu) . \tag{*.1}$$

First-order stationarity is fulfilled :

$$\begin{aligned} E\left(\hat{Z}\right) &= \mu + \alpha_1 E\left(Z_1 - \mu\right) + \alpha_2 E\left(Z_2 - \mu\right) = \mu, \\ E\left(Z_1\right) &= E\left(Z_2\right) = E\left(Z\right) = \mu . \end{aligned} \tag{*.2}$$

The variance should be minimized:

$$\begin{aligned} Var\left(\hat{Z} - Z\right) &= Var\left(\mu + \alpha_1 (Z_1 - \mu) + \alpha_2 (Z_2 - \mu) - Z\right) = Var\left(\alpha_1 (Z_1 - \mu) + \alpha_2 (Z_2 - \mu) - (Z - \mu)\right) = \\ &= E\left(\alpha_1 (Z_1 - \mu) + \alpha_2 (Z_2 - \mu) - (Z - \mu)\right)^2 = \alpha_1^2 Var\left(Z_1\right) + \alpha_2^2 Var\left(Z_2\right) + Var\left(Z\right) + \\ &+ 2\alpha_1 \alpha_2 Cov\left(Z_1, Z_2\right) - 2\alpha_1 Cov\left(Z_1, Z\right) - 2\alpha_2 Cov\left(Z_2, Z\right) = Fct\left(\alpha_1, \alpha_2\right) \underset{\alpha_1, \alpha_2}{\longrightarrow} min . \end{aligned}$$

We now have to minimize a function (*Fct*) of two variables. Other parameters are constant! Thus, we have

$$\begin{cases} \dfrac{\partial}{\partial \alpha_1} Fct\left(\alpha_1, \alpha_2\right) = 0 \\ \dfrac{\partial}{\partial \alpha_2} Fct\left(\alpha_1, \alpha_2\right) = 0 \end{cases} \Rightarrow \begin{cases} 2\alpha_1 Var\left(Z_1\right) + 2\alpha_2 Cov\left(Z_1, Z_2\right) - 2Cov\left(Z_1, Z\right) = 0 \\ 2\alpha_2 Var\left(Z_2\right) + 2\alpha_1 Cov\left(Z_1, Z_2\right) - 2Cov\left(Z_2, Z\right) = 0 \end{cases} \Rightarrow$$

$$\tag{*.3}$$

$$\begin{cases} \alpha_1 Var\left(Z_1\right) + \alpha_2 Cov\left(Z_1, Z_2\right) = Cov\left(Z_1, Z\right) \\ \alpha_2 Var\left(Z_2\right) + \alpha_1 Cov\left(Z_1, Z_2\right) = Cov\left(Z_2, Z\right) \end{cases} \Rightarrow$$

or in matrix form

$$\begin{bmatrix} c_{11} & c_{12} \\ c_{12} & c_{22} \end{bmatrix} \begin{bmatrix} \alpha_1 \\ \alpha_2 \end{bmatrix} = \begin{bmatrix} c_{10} \\ c_{20} \end{bmatrix} \Rightarrow \begin{bmatrix} \alpha_1^L \\ \alpha_2^L \end{bmatrix} = \begin{bmatrix} c_{11} & c_{12} \\ c_{12} & c_{22} \end{bmatrix}^{-1} \begin{bmatrix} c_{10} \\ c_{20} \end{bmatrix} \tag{*.4}$$

As noted above, a model for the covariance function $c(h)$ is given. We calculate all known matrix elements in (*.4) as follows:

$$c_{11} = c_{22} = c(0) \ ,$$
$$c_{12} = c\left(\sqrt{(1.2^2 + 1.1^2)}\right), c_{01} = c\left(\sqrt{(0^2 + 0.2^2)}\right), c_{02} = c\left(\sqrt{(1.2^2 + 0.9^2)}\right)$$

The solution $[\alpha_1^L \ \alpha_2^L]^T$ depends on the model choice. With the weights obtained from (*.4), the estimate corresponds to

$$\hat{z} = \mu + \alpha_1^L (z_1 - \mu) + \alpha_2^L (z_2 - \mu) = 2 + \alpha_1^L (1 - 2) + \alpha_2^L (-3 - 2) = 2 - \alpha_1^L - 5\alpha_2^L .$$
$$(*.5)$$

The kriging variance is calculated by

$$\sigma_{SK}^2 = c_{00} - 2\alpha_1^L c_{01} - 2\alpha_2^L c_{02} . \qquad (*.6)$$

Remark: A model for a covariance function should be based on the empirical covariance function calculated with the given values of the regionalized variable. There are some classes of covariance function models that can be considered. Depending on the covariance model, we can base the model parameters on nonlinear regression methods. But often geostatisticians restrict themselves to an optical fit. Instead of a covariance function we can use another model function—well-known in geostatistics and even more popular—called a *variogram*.

Variogram

Definition 3.1.2-6 Let $\{Z(x)\}, x \in R^n$ be a random field. The following function is called the *variogram* of the random field:

$$\gamma(x,h) = \frac{1}{2} E\{Z(x+h) - Z(x)\}^2 .$$

In contrast to Definition (3.1.2-3) of the covariance function, Definition (3.1.2-6) does not make use of any information about the mean of the corresponding random field. Therefore, this function is interesting for a wide-ranging class of random fields. We assume that this "invisible mean" is identical over the field domain

$$EZ(x+h) = EZ(x).$$

Definition 3.1.2-7 If the variogram of a random field depends only on the length of the translation vector, the field is called *intrinsic stationary*, which means that $\gamma(x,h) = \gamma(|h|)$.

Some properties of the variogram of an intrinsic stationary field are:

1. $\gamma(0) = 0$.
2. $\gamma(h) \geq 0$.

As we are discussing only intrinsic stationary fields in this section, we use the short designation $\gamma(h)$ for $\gamma(|h|)$. Furthermore, if the mean of a random field is known, that is, $EZ(x) = \mu = const$, the following relations between the variogram and co-variance function hold:

$$\gamma(h) = C(o) - C(h)$$
$$C(h) = \gamma(\infty) - \gamma(h)$$

(3-21)

Now we present some variogram models and discuss possible ways to fit them. Obviously, many software tools developed for geostatistics include automatic model-fitting tools. For readers who wish to do more than press a button, we explain how a model can be determined. Fitting the correlation function after fitting the variogram model is simple. Note that the second relation from (3-21) can be applied.

The reality sometimes requires generalizations of existing mathematical models. One such generalization, the so-called *nugget effect*, weakens the variogram property (1) to

$$1+) \; \gamma(0) = 0, \; \gamma(0+) \neq 0$$

(3-22)

The behavior at a very small scale, near the origin of the variogram, is of importance, as it indicates the type of the continuity of the regionalized variable: differentiable, continuous but not differentiable, or discontinuous. If the variogram is not differentiable at the origin, it is a symptom of the nugget effect—a designation that comes from the gold nuggets that are contained in some samples. These nuggets have a natural width that leads to the fact that the values of the variable change abruptly at a very small scale.

We discuss here both kinds of models: without and with a nugget effect, which we denote by *ne*. We introduce the following models: *power variogram, exponential family of variograms, Gauss variogram, spherical variogram,* and *variogram of white noise.* The parameter b is usually called *sill.* From (3-21) it can be seen that the variance of the random field corresponds to $\gamma(\infty)$ because $C(\infty) = 0$. Thus, the variance of a random field without a nugget effect corresponds to the sill b; the variance of fields with a nugget effect is the sum of the sill and this nugget effect, that is, $b + ne$.

Power Variogram

Without a nugget effect (see Fig. 3.8):

$$\gamma(h) = b|h|^p \quad with$$
$$0 < p < 2, \; b > 0$$

(3-23)

With a nugget effect:

$$\gamma(h) = \begin{cases} 0, \; h = 0 \\ ne + b|h|^p, \; h > 0 \end{cases}$$
$$with$$
$$0 < p < 2, \; b, \; ne > 0$$

(3-23')

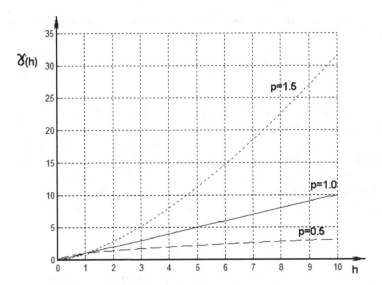

Fig. 3.8 Power variograms without a nugget effect for parameters $p = 0.5$, 1.0, and 1.5; $b = 1$

Remark: For $p = 1$ a power variogram from (3-23) and (3-23′) is also called a *linear variogram*. Equations (3-21) are not true for power variograms because of they increase to infinity.

Exponential Family of Variograms

Without a nugget effect (see Fig. 3.9):

$$\gamma(h) = b - be^{-\frac{|h|^p}{a}}, \quad b, a, p > 0 \tag{3-24}$$

With a nugget effect:

$$\gamma(h) = \begin{cases} 0, h = 0 \\ ne + b\left(1 - e^{-\frac{|h|^p}{a}}\right), h > 0 \end{cases} \tag{3-24′}$$
$$b, a, p > 0$$

Remark: For $p = 1$ a variogram from (3-24) and (3-24′) is called an *exponential variogram*. For $p = 2$ we get a *Gauss variogram*. Equations (3-21) are true and covariance functions exist.

Spherical Variogram

This variogram model is often preferred in practical applications.

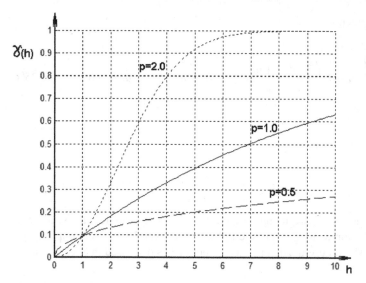

Fig. 3.9 Variograms from the exponential family without a nugget effect for parameters $p = 0.5$, 1.0, and 2.5; $a = 10$; $b = 1$

Without a nugget effect (see Fig. 3.10):

$$\gamma(h) = \begin{cases} b\left(\dfrac{3}{2}\dfrac{|h|}{a} - \dfrac{1}{2}\dfrac{|h|^3}{a^3}\right), & 0 \le |h| \le a \\ b, & |h| > a \end{cases}, \quad a, b > 0 \qquad (3\text{-}25)$$

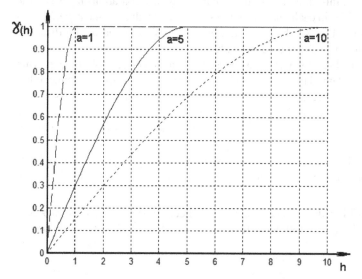

Fig. 3.10 Spherical variograms without nugget effects for parameters $a = 1, 5, 10$; $b = 1$

With a nugget effect:

$$\gamma(h) = \begin{cases} 0, h = 0 \\ ne + b\left(\dfrac{3}{2}\dfrac{|h|}{a} - \dfrac{1}{2}\dfrac{|h|^3}{a^3}\right), & 0 < |h| \le a , \quad a, b > 0 \\ b + ne, & |h| > a \end{cases} \qquad (3\text{-}25')$$

Variogram of White Noise

$$\gamma(h) = \begin{cases} 0, & |h| = 0 \\ b, & |h| > 0 \end{cases} \qquad (3\text{-}26)$$

This is a model of a pure nugget effect. The measurements (z-values) are not correlated. Prediction using a stochastic influence between locations is impossible. In this case one should use the deterministic approaches discussed in Sect. 3.1.1.

Variogram Cloud and Model Fitting

The theoretical definition of the variogram (3.1.2-6) is based on pairs of random values. Variogram fitting starts with evaluating sampled pairs:

$$\gamma^*(h_{ij}) = \frac{1}{2}\left\{z(x_i) - z(x_j)\right\}^2, \quad h_{ij} = |x_i - x_j| \qquad (3\text{-}27)$$

The measurement values are calculated and the resulting dissimilarities $\gamma^*(h_{ij})$ (vertical axis) are plotted against the separation of sampled pairs based on the distances between locations h_{ij} (horizontal axis) forming the *variogram cloud* (see Fig. 3.11a). Further, this cloud is sliced into classes by the separations in space. The average dissimilarities in each class form the sequence of values of the *experimental variogram* shown in Fig. 3.11b. These dissimilarities often increase with distance, as samples near to one another tend to be alike.

The variogram cloud by itself can be seen as a powerful tool for exploring features of spatial data. The distribution of measurements, anomalies, and inhomogeneities can be detected by the way. Looking at the behavior of dissimilarities at short distances, we can make an assumption about nugget effects. In some cases, owing to presence of outliers, the variogram cloud consists of two distinct clouds.

Construction of an experimental variogram is similar to that of the usual *histogram*. We present the definition and show its implementation by a simple example.

Definition 3.1.2-8 Let the distance interval be sliced by $0 \le h_0 < h_1 = h_0 + \Delta < h_0 + 2\Delta < \dots < h_{max}$. The following function is called an *empirical variogram*:

$$\gamma^*(h) = \frac{\sum\limits_{i,\,j=1}^{N} \gamma^*\left(h_{ij} \in \left[h_0 + (k-1)\Delta, h_0 + k\Delta\right]\right)}{\sum\limits_{i,\,j=1}^{N} 1\left(z(x_i),\, z(x_j) : i > j,\, h_{ij} \in \left[h_0 + (k-1)\Delta,\; h_0 + k\Delta\right]\right)}, \qquad (3\text{-}28)$$

$$h_{ij} = \left|x_i - x_j\right|,$$

$$h \in \left[h_0 + (k-1)\Delta,\; h_0 + k\Delta\right]$$

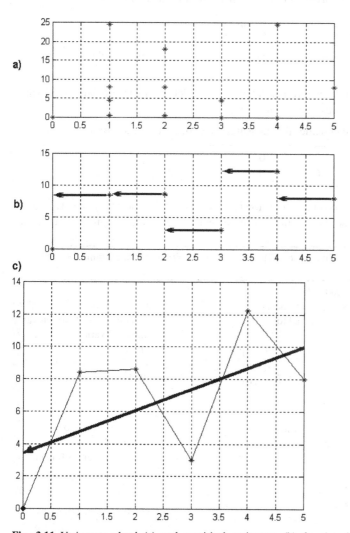

Fig. 3.11 Variogram cloud (a) and empirical variogram (b) for the simple data set from Example 3.1.2.1; (c) linear variogram as variogram model for Example 3.1.2.2

Remark: In connection with Definition (3.1.2-8), we should explain the "hard form" of this term. The denominator counts the number of pairs located in subintervals. The function 1 used there is usual in spatial statistics and indicates the following:

$$1(x) = \begin{cases} 1, \ if \ x \ is \ true \\ 0, \ if \ x \ is \ false \end{cases} \tag{3-29}$$

Definition 3.1.2-9 A variogram model $\hat{\gamma}(h)$ fulfilling the condition

$$|\hat{\gamma}(h) - \gamma^*(h)| \rightarrow \min \tag{3-30}$$

can be used as an *estimator of the true, but unknown variogram* of the random field.

This minimizing procedure is related to all model parameters. Some of the variogram models were presented above.

Example 3.1.2.2 We consider the measurements located over a profile:

x	0.0	1.0	2.0	3.0	4.0	5.0
z	7.0	3.0	6.0	7.0	0.0	3.0

The variogram cloud is related to following six possible distances: $h_{ij} = 0, 1, 2, 3, 4,$ 5. We show stepwise the calculation of variogram cloud and empirical variogram (see Figs. 3.11a, b):

Distance, h_{ij}	Variogram cloud by (3-27)	Empirical variogram by (3-28)
0	Six pairs lead to 0	0
1	Five pairs with coordinates $(0, 1)$, $(1, 2), \ldots (4, 5)$ lead to: $\frac{1}{2}(z(0) - z(1))^2 = \frac{1}{2}(7-3)^2 = 8,$ $\frac{1}{2}(z(1) - z(2))^2 = \frac{1}{2}(3-6)^2 = 4.5,$ $\frac{1}{2}(z(2) - z(3))^2 = \frac{1}{2}(6-7)^2 = 0.5,$ $\frac{1}{2}(z(3) - z(4))^2 = \frac{1}{2}(7-0)^2 = 24.5,$ $\frac{1}{2}(z(4) - z(5))^2 = \frac{1}{2}(0-3)^2 = 4.5.$	The arithmetical mean of the values in the left column leads to the value of the empirical variogram for interval $(0, 1]$: $\gamma^*(h) = \frac{8+4.5+0.5+24.5+4.5}{5} = 8.4,$ $h \in (0, 1]$
2	Four pairs with coordinates $(0, 2)$, $(1, 3), \ldots (3, 5)$ lead to: $\frac{1}{2}(z(0) - z(2))^2 = \frac{1}{2}(7-6)^2 = 0.5,$ $\frac{1}{2}(z(1) - z(3))^2 = \frac{1}{2}(3-7)^2 = 8,$ $\frac{1}{2}(z(2) - z(4))^2 = \frac{1}{2}(6-0)^2 = 18,$ $\frac{1}{2}(z(3) - z(5))^2 = \frac{1}{2}(7-3)^2 = 8.$	The arithmetical mean of values in the left column leads to the value of the empirical variogram for interval $(1, 2]$: $\gamma^*(h) = \frac{0.5+8+18+8}{4} = 8.625,$ $h \in (1, 2]$
3	Three pairs with coordinates $(0, 3)$, $(1, 4)$ und $(2, 5)$ lead to: $\frac{1}{2}(z(0) - z(3))^2 = \frac{1}{2}(7-7)^2 = 0.0,$ $\frac{1}{2}(z(1) - z(4))^2 = \frac{1}{2}(3-0)^2 = 4.5,$ $\frac{1}{2}(z(2) - z(5))^2 = \frac{1}{2}(6-3)^2 = 4.5.$	The arithmetical mean of values in the left column leads to the value of the empirical variogram for interval $(2, 3]$: $\gamma^*(h) = \frac{0.0+4.5+4.5}{3} = 3.0,$ $h \in (2, 3]$

Distance, h_{ij}	Variogram cloud by (3-27)	Empirical variogram by (3-28)
4	Two pairs with coordinates $(0, 4)$ and $(1, 5)$ gives: $\frac{1}{2}(z(0) - z(4))^2 = \frac{1}{2}(7 - 0)^2 = 24.5,$ $\frac{1}{2}(z(1) - z(5))^2 = \frac{1}{2}(3 - 3)^2 = 0.0.$	By evaluation as above we have: $\gamma^*(h) = \dfrac{24.5 + 0.0}{2} = 12.25,$ $h \in (3, 4]$
5	The last pair $(0, 5)$ leads to: $\frac{1}{2}(z(0) - z(5))^2 = \frac{1}{2}(7 - 3)^2 = 8.0.$	$\gamma^*(h) = \dfrac{8}{1} = 8,$ $h \in (4, 5]$

Now we fit the empirical variogram by a variogram model. To simplify matters we choose a linear variogram with a nugget effect [see (3-23′)] for $p = 1$. Thus, we search for parameters ne, b such that

$$|ne + b \cdot h - \gamma^*(h)| \to \min_{ne, b}$$

holds. This is a classical problem of the kind of linear regression discussed in Sect. 3.1.1.

Using (3-9) we get:

$$\binom{b}{ne} = \begin{pmatrix} \sum\limits_{i=1}^{6} h_i^2 & \sum\limits_{i=1}^{6} h_i \\ \sum\limits_{i=1}^{6} h_i & 6 \end{pmatrix}^{-1} \cdot \begin{pmatrix} \sum\limits_{i=1}^{6} \gamma^*(h_i) h_i \\ \sum\limits_{i=1}^{6} \gamma^*(h_i) \end{pmatrix} = \begin{pmatrix} 55 & 15 \\ 15 & 6 \end{pmatrix}^{-1} \cdot \begin{pmatrix} 123.65 \\ 40.275 \end{pmatrix} = \begin{pmatrix} 1.3121 \\ 3.4321 \end{pmatrix} \quad (*.1)$$

Here we set $h_1 = 0$, $h_2 = 1 \dots h_6 = 5$ and use the values of the empirical variogram calculated in the table above. Thus, we fit the following linear variogram model to the empirical variogram (see Fig. 3.11c):

$$\hat{\gamma}(h) = \begin{cases} 0, & h = 0 \\ 3.4321 + 1.3121h, & h > 0 \end{cases} \quad (*.2)$$

It should be noted that this model variogram does not lead to an excellent fit because the nugget effect seems to be too big. However, this is only a teaching example to provide for better understanding of the definitions.

3.1.2.2 Ordinary Kriging

This is the most widely used kriging method not least because of its "realistic" assumptions. Let $\{Z(x)\}$, $x \in R^n$ be an intrinsic stationary random field and $[z(x_1), \dots, z(x_N)]$ the regionalized variable. The mean of this field is unknown, but it is identical over the domain. Thus, it is assumed that

$$EZ(x + h) = EZ(x) = EZ(x_i), \quad i = 1 \dots N \quad (3-31)$$

A prediction of the unknown value at location x_0 should be made, and we use the following form for this estimator:

$$\hat{Z}(x_0) = \sum_{i=1}^{N} \alpha_i Z(x_i) \tag{3-32}$$

The weights from (3-32) have to be determined. An additional requirement is necessary because the mean of estimator (3-32) must be equal to the constant mean of the random field. In statistics such an estimator is called *unbiased*. On the other hand, we have to constrain the weights to sum up to one because in the extreme case when all the measurements are equal to a constant, the estimated value should also be equal to this constant. Thus, we require that

$$\sum_{k=1}^{N} \alpha_k = 1 \tag{3-33}$$

Using (3-31)–(3-33), we can prove that

$$E\hat{Z}(x_0) = E\left(\sum_{i=1}^{N} \alpha_i Z(x_i)\right) = \sum_{i=1}^{N} \alpha_i E(x_i) = EZ(x) \underbrace{\sum_{i=1}^{N} \alpha_i}_{1} = EZ(x) . \tag{3-34}$$

By minimizing the estimation variance we also have to consider the constraint on the weights (3-33):

$$Var\left(Z(x_0) - \hat{Z}(x_0)\right) \rightarrow \min_{\alpha_1 \dots \alpha_N}$$

$$with \tag{3-35}$$

$$\sum_{i=1}^{N} \alpha_i = 1$$

The problem (3-35) can be reformulated by using the so-called *Lagrange coefficient* L as:

$$Var\left(Z(x_0) - \hat{Z}(x_0)\right) - 2L\left(\sum_{i=1}^{N} \alpha_i - 1\right) \rightarrow \min_{\alpha_1 \dots \alpha_N, L} \tag{3-35'}$$

The solution of (3-35') goes through $N+1$ partial derivations:

$$\begin{cases} \dfrac{\partial}{\partial \alpha_k} \sigma_{OK}^2 = 0, & k = 1 \dots N \\[2mm] \dfrac{\partial}{\partial L} \sigma_{OK}^2 = 0 \end{cases}$$

After some simplification this leads to the solution $\left[\alpha_1^L, \dots \alpha_N^L, L^L\right]^T$ of the following LSE in matrix form:

$$
\begin{bmatrix}
\gamma_{11} & \gamma_{12} & \cdots & \gamma_{1N} & 1 \\
\cdots & \cdots & \cdots & \cdots & \cdots \\
\gamma_{N1} & \gamma_{N2} & \cdots & \gamma_{NN} & 1 \\
1 & 1 & \cdots & 1 & 0
\end{bmatrix}
\begin{bmatrix}
\alpha_1^L \\
\cdots \\
\alpha_N^L \\
L^L
\end{bmatrix}
=
\begin{bmatrix}
\gamma_{10} \\
\cdots \\
\gamma_{N0} \\
1
\end{bmatrix}
\quad with \quad \gamma_{ij} = \gamma(h_{ij}), \quad i, j = 0, \ldots, N.
$$

$$(3\text{-}36)$$

The estimated value is calculated with given z-values of the regionalized variable by

$$
z(x_0) = \sum_{k=1}^{N} \alpha_k^L z(x_k) \tag{3-37}
$$

The accuracy of estimation, the kriging variance, can be obtained from (3-35) by

$$
\sigma_{OK}^2 = Var\left(Z(x_0) - \hat{Z}(x_0)\right) = -\gamma(0) - \sum_{i=1}^{N}\sum_{j=1}^{N} \alpha_i^L \alpha_j^L \gamma_{ij} + 2\sum_{i=1}^{N} \alpha_i^L \gamma_{i0}, \quad \gamma_{ij} = \gamma(h_{ij})
$$

$$(3\text{-}38)$$

Remark: Ordinary kriging is also an *exact interpolator* because the estimated value at a given sample point is identical with the z-value at this sample point. According to Sect. 3.1.1 we could say that the interpolation demand is fulfilled.

Example 3.1.2.3 We use again the simple data set from Example 3.1.2.2 and test how ordinary kriging works. The assumption about an exact mean of the random field can be omitted. We only need to know that this mean is constant over the given domain. The regionalized variable takes the following values:

x	0.0	1.0	2.0	3.0	4.0	5.0
z	7.0	3.0	6.0	7.0	0.0	3.0

We want to predict the z-value at point $x_0 = 0.5$ using its closer neighbors $x_1 = 0$, $x_2 = 1$. In other words, we choose the following predictor:

$$
\hat{Z}(x_0) = \sum_{k=1}^{2} \alpha_k Z(x_k) = \alpha_1 Z(x_1) + \alpha_2 Z(x_2), \quad \sum_{k=1}^{2} \alpha_k = \alpha_1 + \alpha_2 = 1 \quad (*.1)
$$

By minimizing the estimation variance (*Fct*) with the constraints on the weights, we obtain:

$$
Var\left(\hat{Z}_0 - Z_0\right) = Var\left(Z(x_0) - \alpha_1 Z(x_1) - \alpha_2 Z(x_2)\right) = -\gamma(0) - \sum_{i=1}^{2}\sum_{j=1}^{2} \alpha_i \alpha_j \gamma_{ij} + 2\sum_{i=1}^{2} \alpha_i \gamma_{i0} \quad (*.2)
$$

$$
= -\gamma(0) - \alpha_1^2 \gamma_{11} - \alpha_2^2 \gamma_{22} - 2\alpha_1 \alpha_2 \gamma_{12} + 2\alpha_1 \gamma_{01} + 2\alpha_2 \gamma_{02} = Fct(\alpha_1, \alpha_2)
$$

$$
with \ \alpha_1 + \alpha_2 = 1 \Rightarrow
$$
$$
Fct(\alpha_1, \alpha_2) - 2L(\alpha_1 + \alpha_2 - 1) \rightarrow \min_{\alpha_1, \alpha_2, L}
$$

We still fit a model variogram (*.2) for the given data from Example (3.1.2.2) and can calculate:

$$\gamma_{11} = \gamma_{22} = 0,$$
$$\gamma_{12} = \gamma(1) = 4.7432,$$
$$\gamma_{01} = \gamma_{02} = \gamma(0.5) = 4.0882$$

Some examples of variogram models and a method for fitting them were discussed above. Equation (*.2) describes a minimizing problem of a function of two variables with an additional constraint on the weights. Using partial derivations with respect to α_1, α_2, L, we get:

$$
\begin{cases}
\dfrac{\partial}{\partial \alpha_1} \{Fct(\alpha_1, \alpha_2) - 2L(\alpha_1 + \alpha_2 - 1)\} = 0 \\[2mm]
\dfrac{\partial}{\partial \alpha_2} \{Fct(\alpha_1, \alpha_2) - 2L(\alpha_1 + \alpha_2 - 1)\} = 0 \;\Rightarrow \\[2mm]
\dfrac{\partial}{\partial L} \{Fct(\alpha_1, \alpha_2) - 2L(\alpha_1 + \alpha_2 - 1)\} = 0
\end{cases}
$$

$$
\begin{cases}
-2\alpha_1\gamma_{11} - 2\alpha_2\gamma_{12} + 2\gamma_{01} - 2L = 0 \\
-2\alpha_2\gamma_{22} - 2\alpha_1\gamma_{12} + 2\gamma_{02} - 2L = 0 \;\Rightarrow \\
\alpha_1 + \alpha_2 - 1 = 0
\end{cases}
$$

$$
\begin{cases}
\alpha_1\gamma_{11} + \alpha_2\gamma_{12} + L = \gamma_{01} \\
\alpha_2\gamma_{22} + \alpha_1\gamma_{12} + L = \gamma_{02} \;\Rightarrow \\
\alpha_1 + \alpha_2 - 1 = 0
\end{cases}
\begin{bmatrix}
0 & 4.7432 & 1 \\
4.7432 & 0 & 1 \\
1 & 1 & 0
\end{bmatrix}
\begin{bmatrix}
\alpha_1^L \\
\alpha_2^L \\
L^L
\end{bmatrix}
=
\begin{bmatrix}
4.0882 \\
4.0882 \\
1
\end{bmatrix}
$$

The solution $\begin{bmatrix} \alpha_1^L & \alpha_2^L & L \end{bmatrix}^T = [0.5, 0.5, 1.716]$ depends strongly on the fitted variogram model. After setting all the parameters we get the estimated z-value at point $x_0 = 0.5$ from:

$$z(0.5) = \alpha_1^L z_1 + \alpha_2^L z_2 = 0.5 \cdot 7 + 0.5 \cdot 3 = 5$$

The kriging variance can be obtained from (*.2) by using the given parameters:

$$
\begin{aligned}
\sigma_{OK}^2(0.5) &= -\gamma(0) - \left[\alpha_1^L\right]^2 \gamma_{11} - \left[\alpha_2^L\right]^2 \gamma_{22} - 2\alpha_1^L\alpha_2^L\gamma_{12} + 2\alpha_1^L\gamma_{01} + 2\alpha_2^L\gamma_{02} \\
&= -2\alpha_1^L\alpha_2^L\gamma(1) + 2\alpha_1^L\gamma(0.5) + 2\alpha_2^L\gamma(0.5) \\
&= -2 \cdot 0.5 \cdot 0.5 \cdot 4.7443 + 2 \cdot 0.5 \cdot 4.0882 \cdot 2 = 5.8042
\end{aligned}
$$

3.1.2.3 Universal Kriging

Universal kriging is a spatial multiple regression method implementing a model that splits the random field into two parts. The first corresponds to a linear combination of $(M+1)$ deterministic functions that are known at any point in the region. The second part is a random component that is called the *residual random function*. This generalized model needs additional specifications.

Let $\{Z(x)\}, x \in R^n$ be a random field that can be divided into two parts: a deterministic part often called the *drift* and a stochastic part with mean zero $[ES(x) = 0]$:

$$Z(x) = m(x) + S(x) \tag{3-39}$$

The drift or the mean of this field can be presented as the following sum:

$$m(x) = EZ(x) = \sum_{l=0}^{M} a_l f_l(x),$$
$$a_l \neq 0, \ f_0 = 1, l = 0, ..., M, \tag{3-40}$$

and let $[z(x_1), ..., z(x_N)]$ be a regionalized variable. We estimate an unknown value at the point x_0 using the following form of the predictor:

$$\hat{Z}(x_0) = \sum_{k=1}^{N} \alpha_k Z(x_k) \tag{3-41}$$

We are looking for an unbiased estimate; that is, the mean of the difference between the predictor from (3-41) and the true but unknown random variable at point x_0 should be equal to zero:

$$E\left(\hat{Z}(x_0) - Z(x_0)\right) = 0 \Rightarrow \sum_{i=1}^{N} \alpha_i \sum_{l=0}^{M} f_l(x_i) = E\hat{Z}(x_0) = EZ(x_0)$$
$$= \sum_{l=0}^{M} f_l(x_0) \Rightarrow \sum_{i=1}^{N} \alpha_i f_l(x_i) = f_l(x_0), \quad l = 0, ..., M \tag{3-42}$$

For the constant function $f_0(x)$ this is the condition with which we are familiar from ordinary kriging:

$$\sum_{k=1}^{N} \alpha_k = 1 \tag{3-42'}$$

Developing the expression for the estimation variance, introducing the constraints into the objective function together with Lagrange's parameters $L_0, ..., L_M$, and minimizing, we obtain

$$Var\left(\hat{Z}(x_0) - Z(x_0)\right) + 2\sum_{l=0}^{M} L_l\left(\sum_{i=1}^{N} \alpha_i f_l(x_i) - f_l(x_0)\right) \to \min(\alpha_1 ... \alpha_N, L_0 ... L_M)$$

and the following LSE:

$$\begin{cases} \sum_{j=1}^{N} \alpha_j c_{ij} - \sum_{l=0}^{M} L_l f_l(x_i) = c_{i0} \quad i = 1, ..., N \\ \sum_{j=1}^{N} \alpha_j f_l(x_j) = f_l(x_0), \quad l = 0, ..., M \\ c_{ij} = Cov(Z(x_i), Z(x_j)), \ i, j = 0, ..., N \end{cases} \tag{3-43}$$

and in matrix notation:

$$\begin{bmatrix} C & F \\ F^T & 0 \end{bmatrix} \cdot \begin{bmatrix} \alpha^L \\ L^L \end{bmatrix} = \begin{bmatrix} c_0 \\ f^T \end{bmatrix}$$

with

$$F(i, \cdot) = [f_0(x_i), \dots, f_M(x_i)], \quad i = 1, \dots, N$$

$$f = [f_0(x_0), \dots, f_M(x_0)]^T$$

(3-43')

After solving (3-43') by inverting the LSE matrix, we obtain weights as part of the solution vector $[\alpha^L, L^L]^T$. Thus, the prediction at point x_0 can be presented using the given z-values of the regionalized variable:

$$\hat{z}(x_0) = \sum_{k=1}^{N} \alpha_k^L z(x_k)$$

(3-44)

The accuracy of this prediction is given by the universal kriging variance:

$$\sigma_{UK}^2 = Var\left(Z(x_0) - \hat{Z}(x_0)\right) = c_{00} + \sum_{k,j=1}^{N} \alpha_j^L \alpha_k^L c_{kj} - 2 \sum_{k=1}^{N} \alpha_k^L c_{0k}$$

(3-45)

More about different kriging methods and their generalizations can be found in Wackernagel (1995).

3.1.2.4 Cross Validation and Goodness Measure

Following is a brief overview of the method of cross validation and the goodness measure for the evaluation of the choice of a variogram model.

We assume that all the modeling steps are finished: the variogram cloud and the empirical variogram are calculated based on given measurements, a model for the empirical variogram is fitted, a kriging approach is applied, and unknown values with corresponding kriging variances are obtained. Is the chosen model variogram now really good or bad for this case? Is there any way to answer this question?

Cross validation, which tests the empirical accuracy of kriging is such a way, and it works as follows. From N known z-values of the regionalized variable, a value is chosen and has to be kriged by the remaining values, using the chosen variogram model. This procedure is applied stepwise for each z-value. First we prove that

$$G_1 = \frac{1}{N} \sum_{i=1}^{N} (z(x_i) - \hat{Z}(x_i)) \approx 0$$

(3-46)

If (3-46) is false, there are systematic over- or underestimation effects, thus indicating that the chosen model is not really perfect.

Second we calculate and prove that

$$G_2 = \frac{1}{N} \sum_{i=1}^{N} \frac{(z(x_i) - \hat{Z}(x_i))^2}{\sigma_{[i]}^2} \approx 1$$

(3-47)

The denominators $\sigma^2_{[i]}$ in (3-47) describe the obtained kriging variance. In (3-47) the theoretical and the experimental accuracy of the estimated values can be compared. If both (3-46) and (3-47) hold, the chosen variogram model would seem to be acceptable.

3.2 Describing Seeming "Chaos" in Measurements by an Analytical Function

Obviously, it is very tempting to try to obtain a single "world formula," one that can describe all known processes and all spatial and temporal structures. By nature human beings tend to cause their own chaos, but by nature they also try to organize chaos caused by someone else.

There are many mathematical approaches that help us recognize data structures. We touched on the basic idea of simple regression methods and splines in Sect. 3.1.1. Here, in Sect. 3.2.1 we discuss some common principles of different surface approximation methods and in Sect. 3.2.2 we present some stochastic models.

3.2.1 Different Regression Approaches: Basic Ideas and Ways for Further Generalizations

There are many different reasons and objectives for fitting measurements by an analytical curve or by a surface—for example, parameter estimation, data smoothing, functional representation, and data reduction—and we discuss each of these reasons separately. *Parameter estimation* is used if the form of an analytical function describing a process is dictated by parameters that have a specific physical meaning. The goal is then to estimate those parameters as accurately as possible from the given measurements. *Data smoothing* by an analytical function helps to reduce measurement errors. We can hope that with the analytical function fitting a process these errors will be more or less smoothed out. *Functional representation* of a discrete set of measurements may have a number of advantages. First, values at any point in the range of representation can be predicted (see Sect. 3.1). Second, the functional approximation can be used for a deeper analysis of the data, for example, for determining derivatives, definite integrals, and so on. *Data reduction* is necessary if the given data set is too big for further analysis. In particular, this means that numerical problems can occur. We can approximate the given data set by an analytical function that has fewer parameters than the number of measurements that are given.

There are many approaches that lead to functional approximations owing to various smoothing criteria, which are dictated by real problems [see Dierckx (1993)]. The least-squares criterion is the one used most often. Sometimes there are constraints that lead to further generalizations of existing methods. We present some of these that, on the one hand, are useful and, on the other, are simple to implement, omitting detailed discussions about the numerical optimization of the corresponding algorithms.

We start with the *least-squares criterion* of smoothing and discuss its widespread applicability. Let $[z(x_1), \ldots, z(x_N)] = [z_1, \ldots, z_N]$ be measurements at locations $[x_1, \ldots, x_N]$. An analytical function $f(x)$ must be found that fulfills the following least-squares criterion:

$$\sum_{k=1}^{N} (f(x_i) - z_i)^2 \rightarrow \min \qquad (3\text{-}48)$$

We denote all coordinates of the locations by x. Usually, the function f is assumed to depend on $M + 1$ parameters, which should be "optimally" chosen so that the following holds:

$$\sum_{\substack{k=1 \\ M \leq N-1}}^{N} (f(a_0, \ldots a_M, x_i) - z_i)^2 \rightarrow \min(a_0, \ldots, a_M), \qquad (3\text{-}48')$$

Parameter estimation means determining $a_0, \ldots, a_M$. A single function class or many different ones be considered. Data smoothing and functional representation lead to the choice of an analytical function f describing the data structure of interest. If the interpolation demand is fulfilled, $M = N - 1$. In this case the z-values of given measurements are reconstructed at their locations. For data reduction the number M of chosen parameters has to be less than $N - 1$ by some order.

The solution of the minimizing problem (3-48') *without any additional constraints* goes through

$$\frac{\partial}{\partial a_i} \left[\sum_{k=1}^{N} (f(a_0, \ldots, a_M, x_k) - z_k)^2 \right] = 0, \; i = 0 \ldots M \qquad (3\text{-}49)$$

The resulting system of equations corresponds to

$$\sum_{k=1}^{N} \left(f(a_0, \ldots, a_M, x_k) \cdot \frac{\partial}{\partial a_i} [f(a_0, \ldots, a_M, x_k)] \right)$$
$$= \sum_{k=1}^{N} z_k \cdot \frac{\partial}{\partial a_i} [f(a_0, \ldots a_M, x_k)], i = 0 \ldots M \qquad (3\text{-}49')$$

and is not always linear! For example, it is a linear system of equations (LSE) if

$$\frac{\partial}{\partial a_i} [f(a_0, \ldots, a_M, x_k)] = g(x_k), \; k = 1 \ldots N, \; i = 1 \ldots M$$

Thus, the following form of analytical functions f can be recommended:

$$f(a_0, \ldots, a_M, x) = \sum_{i=0}^{M} a_i g^i(x) \qquad (3\text{-}50)$$

The functions $g^i(x)$, $i = 0 \ldots M$ should be chosen carefully. First, these functions should have some practical sense. For example, if the data are expected to have a wave structure, it is better to use a trigonometric function rather than a linear one. Second, the resulting matrix of the LSE has to be regular. Unfortunately, we cannot give any general rules here, so intuition and experience are essential.

Before solving, all possible constraints in the special form $c_j(a_0, \ldots a_M) = 0$, $j = 1 \ldots K$ should be cleared. There are generally certain practical reasons for these constraints. Their number K should be restricted in relation to the number N of measurements and to the number $M + 1$ of unknown parameters:

$$K + M + 1 \leq N \qquad (3\text{-}50')$$

The minimizing problem *with K additional constraints* corresponds to

$$\sum_{k=1}^{N} (f(a_0, \ldots a_M, x_i) - z_i)^2 + \sum_{j=1}^{K} L_j \cdot c_j(a_0, \ldots a_M)$$

$$\rightarrow \min(a_0, \ldots, a_M, L_1, \ldots, L_K), \ K + M + 1 \leq N \qquad (3\text{-}51)$$

and results in the following system of equations:

$$\begin{cases} \dfrac{\partial}{\partial a_i} \left[\sum_{k=1}^{N} (f(a_0, \ldots a_M, x_i) - z_i)^2 + \sum_{j=1}^{K} L_j \cdot c_j(a_0, \ldots a_M) \right] = 0 \\[4mm] \dfrac{\partial}{\partial L_m} \left[\sum_{j=1}^{K} L_j \cdot c_j(a_0, \ldots a_M) \right] = 0 \Rightarrow c_m(a_0, \ldots a_M) = 0 \\[2mm] i = 0 \ldots M, \ m = 1 \ldots K \end{cases} \qquad (3\text{-}51')$$

where the L_j denote Lagrange parameters. Again, the generalized system of equations from (3-51) can be nonlinear.

Remark: If one cannot avoid using function such as $f(a_0, \ldots a_M, x)$ and constraints $c_j(a_0, \ldots a_M)$, $j = 1 \ldots K$, which lead to nonlinear systems of equations, some special techniques for searching for a minimum in (3-51) should be employed. Generally, these methods do not provide an exact solution, but do yield a numerical approximation. For example, the well-known down-hill-simplex algorithm can be used for implementation.

Now, let us consider a simple example in order to demonstrate various regression models.

Example 3.2.1.1 We consider the following temporal measurements from Example 3.1.1.1:

$$z_1 = z(1) = 0.1, \quad z_2 = z(4) = 0.2, \quad z_3 = z(7) = -0.1, \quad z_4 = z(10) = -0.2$$

In (*.4) of Example 3.1.1.1'''' a linear function fitting these measurements was calculated and is given by:

$$z(x) = -0.04x + 0.22 \tag{*.1}$$

Equation (*.1) is an analytical function obtained by a linear regression approach without any constraints. It is a possible mathematical model to help describe the structure of the data that considers all the measurements in the same way. What generalized mathematical model can fit these measurements for the case, if the second measurement is given exactly and should be retained by the linear regression? It means that we are searching for two parameters a and b such that

$$\begin{aligned} F(a, b) &= \sum_{i=1}^{4} (ax_i + b - z_i)^2 \\ &= (a \cdot 1 + b - 0.1)^2 + (a \cdot 4 + b - 0.2)^2 + (a \cdot 7 + b - (-0.1))^2 \\ &\quad + (a \cdot 10 + b - (-0.2))^2 \to \min_{a,\, b} \end{aligned} \tag{*.2}$$

with the additional constraint

$$a \cdot 4 + b = 0.2 \tag{*.2'}$$

The generalized model, a linear regression approach with a fixed point, corresponds to

$$F(a, b, L) = \sum_{i=1}^{4} (ax_i + b - z_i)^2 + 2L \cdot (a \cdot 4 + b - 0.2) \to \min_{a,\, b,\, L}$$

The Lagrange parameter, which is equal to $2L$, is used to further simplify the equations. Considering the partial derivates leads to

$$\begin{cases} \dfrac{\partial F}{\partial a} = \sum\limits_{i=1}^{4} 2(ax_i + b - z_i)x_i + 2L \cdot 4 = 0 \\[2mm] \dfrac{\partial F}{\partial b} = \sum\limits_{i=1}^{4} 2(ax_i + b - z_i) \cdot 1 + 2 \cdot L = 0 \\[2mm] \dfrac{\partial F}{\partial b} = 2(a \cdot 4 + b - 0.2) = 0 \end{cases} \Rightarrow \begin{cases} \sum\limits_{i=1}^{4}(ax_i + b - z_i)x_i + 4L = 0 \\[2mm] \sum\limits_{i=1}^{4}(ax_i + b - z_i) \cdot 1 + L = 0 \\[2mm] 4a + b = 0.2 \end{cases} \Rightarrow \tag{*.3}$$

In matrix form we have

$$\begin{pmatrix} \sum\limits_{i=1}^{4} x_i^2 & \sum\limits_{i=1}^{4} x_i & 4 \\ \sum\limits_{i=1}^{4} x_i & 4 & 1 \\ 4 & 1 & 0 \end{pmatrix} \begin{pmatrix} a \\ b \\ L \end{pmatrix} = \begin{pmatrix} \sum\limits_{i=1}^{4} x_i z_i \\ \sum\limits_{i=1}^{4} z_i \\ 0.2 \end{pmatrix} \Rightarrow \begin{pmatrix} a^L \\ b^L \\ L^L \end{pmatrix} = \begin{pmatrix} 166 & 22 & 4 \\ 22 & 4 & 1 \\ 4 & 1 & 0 \end{pmatrix}^{-1} \begin{pmatrix} -1.8 \\ 0.0 \\ 0.2 \end{pmatrix} = \begin{pmatrix} -0.0556 \\ 0.4222 \\ -0.4667 \end{pmatrix}. \tag{*.3'}$$

Now we obtain an alternative linear function fitting the given measurements by

$$z(x) = -0.0556x + 0.4222 \qquad (*.4)$$

The changed conditions – namely using the additional constraint $(*.2')$ - lead to a changed mathematical model and result in another linear structure $(*.4)$ in comparison with $(*.1)$. It can be proved that the demand $(*.2')$ is fulfilled. Thus, the fitted regression line goes through the second point $x = 4$ and retains the z-value 0.2 (see Fig. 3.12).

In order to demonstrate the relative freedom of choosing functions $g^i(x)$, $i = 0 \ldots M$ from (3-50), we approximate the same data set again using the following analytical function without further constraints:

$$f(a, b, x) = ax + b \cdot \cos(\pi x) \qquad (*.5)$$

We must determine two parameters a and b that minimize $F(a, b)$:

$$F(a, b) = \sum_{i=1}^{4} (ax_i + b \cdot \cos(\pi x_i) - z_i)^2 \to \min_{a, b} \qquad (*.5')$$

Considering partial derivatives, we get

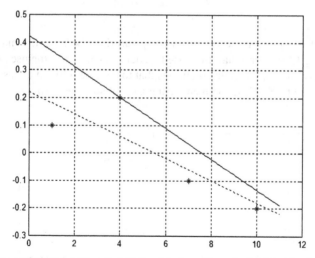

Fig. 3.12 Two analytical functions that fit the data from Example 3.2.1.1. The dotted line corresponds to the linear regression approach without constraints $(*.1)$. The solid line describes the linear structure with a fixed second point (linear regression approach with constraints)

$$\begin{cases} \dfrac{\partial F}{\partial a} = \sum_{i=1}^{4} 2\,(ax_i + b\cos(\pi x_i) - z_i)\,x_i = 0 \\[2mm] \dfrac{\partial F}{\partial b} = \sum_{i=1}^{4} 2\,(ax_i + b\cos(\pi x_i) - z_i) \cdot \cos(\pi x) = 0 \end{cases} \Rightarrow$$

(*.6)

$$\begin{pmatrix} \sum_{i=1}^{4} x_i^2 & \sum_{i=1}^{4} x_i \cos(\pi x_i) \\[2mm] \sum_{i=1}^{4} x_i \cos(\pi x_i) & \sum_{i=1}^{4} \cos^2(\pi x_i) \end{pmatrix} \begin{pmatrix} a \\ b \end{pmatrix} = \begin{pmatrix} \sum_{i=1}^{4} x_i z_i \\[2mm] \sum_{i=1}^{4} z_i \cos(\pi x_i) \end{pmatrix}$$

Using the given values and solving the equation system, we obtain

$$\begin{pmatrix} a^L \\ b^L \end{pmatrix} = \begin{pmatrix} 166\ 6.0 \\ 6.0\ 4.0 \end{pmatrix}^{-1} \begin{pmatrix} -1.8 \\ 0.0 \end{pmatrix} = \begin{pmatrix} -0.0115 \\ 0.0172 \end{pmatrix}$$

(*.6′)

This linear regression approach leads to the analytical function shown in Fig. 3.13:

$$f(x) = -0.0115x + 0.0172 \cdot \cos(\pi x)$$

(*.7)

Obviously, the model expressed in (*.5) is far from perfect. But it is meant as an instructional tool: If someone carelessly uses function

$$f(a, b, x) = ax + b \cdot \sin(\pi x)$$

(*.8)

instead of (*.5) he or she is unpleasantly surprised because the matrix of the corresponding LSE is irregular. Interested readers can convince themselves.

Fortunately, there are certain conditions that to have be fulfilled to ensure a regular matrix of the LSE. We recommend using a set of so-called *orthogonal functions*

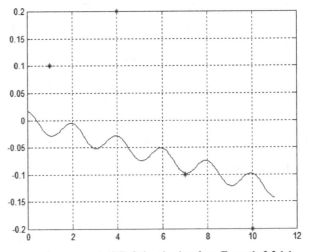

Fig. 3.13 Linear regression approach (*.7) fitting the data from Example 3.2.1.1

$g^i(x)$, $i = 0 \ldots M$, which we discuss below. The location of the measurements is of importance. Generally, this is a nontrivial topic, which is touched on in Dierckx (1993).

Finally, we generalize the model (*.5) to

$$f(a, b, x) = ax + b \cdot \cos(\pi x) + c \qquad (*.9)$$

Considering partial derivations with respect to a, b, and c in order to minimize the sum of squared differences, we obtain

$$\begin{pmatrix} \sum_{i=1}^{4} x_i^2 & \sum_{i=1}^{4} x_i \cos(\pi x_i) & \sum_{i=1}^{4} x_i \\ \sum_{i=1}^{4} x_i \cos(\pi x_i) & \sum_{i=1}^{4} \cos^2(\pi x_i) & \sum_{i=1}^{4} \cos(\pi x_i) \\ \sum_{i=1}^{4} x_i & \sum_{i=1}^{4} \cos(\pi x_i) & 4 \end{pmatrix} \begin{pmatrix} a \\ b \\ c \end{pmatrix} = \begin{pmatrix} \sum_{i=1}^{4} z_i x_i \\ \sum_{i=1}^{4} z_i \cos(\pi x_i) \\ \sum_{i=1}^{4} z_i \end{pmatrix}$$

$$(*.10)$$

Using the given values and inverting the matrix leads to

$$\begin{pmatrix} a \\ b \\ c \end{pmatrix} = \begin{pmatrix} 166 & 6 & 22 \\ 6 & 4 & 0 \\ 22 & 0 & 4 \end{pmatrix}^{-1} \begin{pmatrix} -1.8 \\ 0.0 \\ 0.0 \end{pmatrix} = \begin{pmatrix} -0.05 \\ 0.075 \\ 0.275 \end{pmatrix} \qquad (*.10')$$

Figure 3.14 shows the following analytical function that fits the data:

$$f(x) = -0.05x + 0.075 \cdot \cos(\pi x) + 0.275 \qquad (*.11)$$

The question of which analytical function fits these data best is a philosophical one. There are many functions that fit data best with respect to some optimization criterion. Obviously, different constraints can be assumed and the real applications of the chosen mathematical model play an important role.

If we compare the four models that have been considered here, we should look at the sums of squared differences, given as follows:

- Linear model without constraints leads to $Sum = 0.028$.
- Linear model with fixed second point leads to $Sum = 0.0931$.
- Linear model (*.7) leads to $Sum = 0.0794$.
- Linear model (*.11) leads to $Sum = 0.01$.

Therefore, the linear model (*.11) is optimal one with respect to the sum of squares. Perhaps one of our readers can find a better model!

3.2.1.1 Orthogonal Functions and the Gram-Schmidt Method

Orthogonal polynomials, especially so-called the *Legendre polynomials* are widespread in geodesy. A special kind of orthogonal function is given by what are known as *wavelets*. Using approximation with orthogonal functions leads to a simpler calculation of the following integral, which is replaced by a sum:

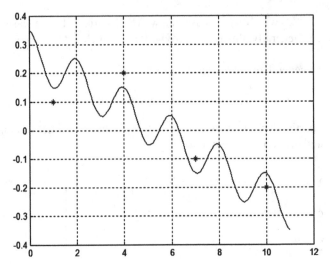

Fig. 3.14 Linear regression approach (*.11) fitting the data from Example 3.2.1.1

$$\int_a^b f(x)\,g(x)\,dx = \langle f(x),\, g(x)\rangle = \left\langle \sum_{k=0}^{\infty} \alpha_k^f e_k(x),\, \sum_{k=0}^{\infty} \alpha_k^g e_k(x)\right\rangle = \sum_{k=0}^{\infty} \alpha_k^f \alpha_k^g \|e_k\|^2,$$

$$\langle e_k,\, e_j\rangle = \begin{cases} 0, & k \neq j \\ \|e_k\|^2, & k = j \end{cases}$$

$$\alpha_k^f = \langle f(x),\, e_k(x)\rangle,\quad \alpha_k^g = \langle g(x),\, e_k(x)\rangle$$

$$(3\text{-}52)$$

Definition 3.2.1-1 Let u, v be two regular functions defined for an interval $[a, b]$. Further, let w be a nonnegative function defined for the same interval. The following value is called the *scalar product of the functions u, v* with respect to the weight function w:

$$\langle u,\, v\rangle_w = \int_a^b u(x)\cdot v(x)\cdot w(x)\,dx \qquad (3\text{-}53)$$

Example 3.2.1.2 Calculation of the scalar product for following functions:

$$w(x) = x^2$$
$$v(x) = x$$
$$u(x) = 3x + 5$$

$$\langle u,\, v\rangle_w = \int_a^b (3x+5)\cdot x\cdot x^2 dx = \frac{3}{5}\left(b^5 - a^5\right) + \frac{5}{4}\left(b^4 - a^4\right).$$

Definition 3.2.1-2 Two regular functions u, v defined for the interval $[a, b]$ are called *orthogonal* if their scalar product is equal to zero.

The importance of orthogonal functions in functional spaces is similar to that of orthogonal basis vectors in Euclidean space. Analogously to the representation of any vector as the sum of basis vectors, any function $f(x)$ defined on interval $[a, b]$ can be approximated by using the orthogonal functions $\{e_i(x)\}_{i=1}^{\infty}$:

$$f(x) = \sum_{i=1}^{\infty} f_i e_i(x)$$

$$f_i = \frac{\langle f(x), e_i(x)\rangle}{\langle e_i(x), e_i(x)\rangle}, \quad i = 1, 2, \ldots .$$

(3-54)

Example 3.2.1.3 Are the polynomials 1, x, x^2, ... orthogonal in relation to the interval $[a, b]$? We can answer this question using Definitions 3.2.1-1 and 3.2.1-2:

$$\{g_i(x) = x^{i-1}\}_{i=1}^{\infty}: \quad \forall a < b:$$

$$\langle g_i, g_j\rangle = \int_a^b x^{i+j-2}dx = \frac{1}{i+j-1}\left(b^{i+j-1} - a^{i+j-1}\right) \neq 0$$

Thus, these polynomials are not orthogonal, but linear independent, which means that it is not possible to represent such a polynomial as a weighted sum or a linear combination of the other polynomials:

$$g_i \neq \sum_{k=0, k\neq i}^{\infty} \alpha_k g_k(x) \quad \forall \alpha_k, k = 0\ldots$$

$$i = 0, \ldots$$

There is a method generating a system of orthogonal functions from a linear-independent system of functions called Gram-Schmidt orthogonalization:

Let $\{g_1(x), g_2(x), \ldots g_n(x), \ldots\}$ be linear-independent functions on the interval $[a, b]$. The functions $\{e_1(x), e_2(x), \ldots, e_n(x), \ldots\}$, based on $\{g_1(x), g_2(x), \ldots g_n(x), \ldots\}$, build an orthogonal basis on the interval $[a, b]$:

$e_1(x) = g_1(x),$

$e_2(x) = g_2(x) - c_{1,2}\dfrac{e_1(x)}{\sqrt{\langle e_1(x), e_1(x)\rangle}}, \ldots$

$e_n(x) = g_n(x) - c_{1,n}\dfrac{e_1(x)}{\sqrt{\langle e_1(x), e_1(x)\rangle}} - c_{2,n}\dfrac{e_2(x)}{\sqrt{\langle e_2(x), e_2(x)\rangle}} - \ldots - c_{n-1,n}\dfrac{e_{n-1}(x)}{\sqrt{\langle e_{n-1}(x), e_{n-1}(x)\rangle}}, \ldots$

The constants used here can be calculated by

$$c_{j,k} = \frac{\langle e_j(x), g_k(x)\rangle}{\sqrt{\langle e_j(x), e_j(x)\rangle}}, \quad j \leq k-1$$

(3-55)

Let us consider Example 3.2.1.3′ to prove the Gram-Schmidt method.

Example 3.2.1.3′ The following system $\{g_i(x) = x^{i-1}\}_{i=1}^{\infty}$ has to be orthogonalized on $[-1, 1]$. Using (3-55) we get successively:

$$e_1(x) = 1,$$

$$e_2(x) = x \quad with \quad c_{1,2} \propto \int_{-1}^{1} 1 \cdot x\, dx = 0 \tag{*.1}$$

$$e_3(x) = x^2 - \frac{1}{3} \quad with \quad c_{2,3} \propto \int_{-1}^{1} x \cdot x^2\, dx = 0 \quad and \quad c_{1,3} = \frac{\int_{-1}^{1} 1 \cdot x^2\, dx}{\sqrt{\int_{-1}^{1} 1 \cdot 1\, dx}} = \frac{2/3}{\sqrt{2}} = \frac{\sqrt{2}}{3}$$

etc.

$$\tag{*.2}$$

Figure 3.15 shows the first three functions of this orthogonal basis on the interval $[-1, 1]$.

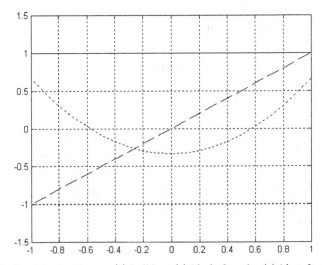

Fig. 3.15 The first three functions, $e_1(x)$ (*solid*), $e_2(x)$ (*dashed*), and $e_3(x)$ (*dotted*), of the orthogonal basis from Example 3.2.1.3′

Example 3.2.1.4 Let us prove that the system

$$1, \cos(x), \sin(x), \cos(2x), \sin(2x), \ldots \cos(nx), \sin(nx)$$

is orthogonal on $[-\pi, \pi]$. We can first prove the following:

$$\int_{-\pi}^{\pi} \cos(jx)\cos(kx)\,dx = \begin{cases} 0, & j \neq k \\ 2\pi, & j = k = 0 \\ \pi, & j = k > 0 \end{cases} \qquad (*.1)$$

$$\int_{-\pi}^{\pi} \sin(jx)\sin(kx)\,dx = \begin{cases} 0, & j \neq k,\ j,\ k > 0 \\ \pi, & j = k > 0 \end{cases} \qquad (*.2)$$

$$\int_{-\pi}^{\pi} \cos(jx)\sin(kx)\,dx = 0, \quad \forall j \geq 0,\ k > 0 \qquad (*.3)$$

For (*.1) and (*.2) the following well-known trigonometric relations can be used:

$$\int_{-\pi}^{\pi} \cos(jx)\cos(kx)\,dx = \frac{1}{2}\int_{-\pi}^{\pi} [\cos([j+k]x) + \cos([j-k]x)]\,dx$$

and

$$\int_{-\pi}^{\pi} \sin(jx)\sin(kx)\,dx = \frac{1}{2}\int_{-\pi}^{\pi} [\cos([j+k]x) - \cos([j-k]x)]\,dx$$

For the integral in (*.3),

$$\int_{-\pi}^{0} \cos(jx)\sin(kx)\,dx = -\int_{0}^{\pi} \cos(jx)\sin(kx)\,dx \Rightarrow \int_{-\pi}^{\pi} \cos(jx)\sin(kx)\,dx = 0$$

holds. This fact taken into account in Example 3.2.1.4 is used to construct a functional approximation by a *Fourier transform*, an approximation that is meaningful for data characterized by a wave structure. We review this topic briefly; more details can be found in Bracewell (1978) and Hamming (1973).

3.2.1.2 Approximation with the Fourier Transform (1D)

We restrict ourselves here to the one-dimensional case. Measurements $[z(x_1), \ldots, z(x_N)] = [z_1, \ldots, z_N]$ from an unknown function $z(x)$ should be approximated in $[-\pi, \pi]$ by the following analytic function:

$$f_n(x) = \frac{1}{2}a_0 + \sum_{k=1}^{n} [a_k \cos(kx) + b_k \sin(kx)] \qquad (3\text{-}56)$$

Using (3-53) and (3-54), we can represent this function as

$$z(x) = \sum_{i=1}^{\infty} c_i e_i(x) = \frac{1}{2}a_0 + \sum_{k=1}^{\infty} [a_k \cos(kx) + b_k \sin(kx)] \approx \frac{1}{2}a_0 + \sum_{k=1}^{n} a_k \cos(kx) + \sum_{k=1}^{n} b_k \sin(kx)$$

$$c_i = \frac{\langle z(x), e_i(x) \rangle}{\langle e_i(x), e_i(x) \rangle}, \quad i = 1, 2, \ldots . \Rightarrow$$

$$(3\text{-}57)$$

$$\langle z(x), 1 \rangle = \int_{-\pi}^{\pi} z(x)\,dx, \ \langle 1, 1 \rangle = 2\pi \Rightarrow a_0 = \frac{1}{\pi} \int_{-\pi}^{\pi} z(x)\,dx,$$

$$\langle z(x), \cos(kx) \rangle = \int_{-\pi}^{\pi} z(x) \cos(kx)\,dx, \ \langle \cos(kx), \cos(kx) \rangle = \pi \Rightarrow a_k = \frac{1}{\pi} \int_{-\pi}^{\pi} z(x) \cos(kx)\,dx,$$

$$\langle z(x), \sin(kx) \rangle = \int_{-\pi}^{\pi} z(x) \sin(kx)\,dx, \ \langle \sin(kx), \sin(kx) \rangle = \pi \Rightarrow b_k = \frac{1}{\pi} \int_{-\pi}^{\pi} z(x) \sin(kx)\,dx$$

$$(3\text{-}57')$$

Here we use relations (*.1)–(*.3) from Example 3.2.1.4. As shown above, we obtain equations for calculating coefficients where the coefficient is given as an integral and depends on the continuous but unknown function $z(x)$. Thus, these equations should also be "translated" into a discrete form based on the given measurements $[z(x_1), \ldots, z(x_N)] = [z_1, \ldots, z_N]$. This can be done by applying the well-known equation (*trapezium rule*):

$$\int_a^b f(x)\,dx \approx \frac{b-a}{2}(f(a) + f(b)) \tag{3-58}$$

If the interval $[-\pi, \pi]$ is divided into $N-1$ subintervals with knots $-\pi = x_1$, $x_2, \ldots, x_N = \pi$ equation (3-58) can be applied for each interval separately. The final sum leads to

$$\int_a^b f(x)\,dx \approx \frac{1}{2} \sum_{i=1}^{N-1} (f(x_i) + f(x_{i+1}))(x_{i+1} - x_i) \tag{3-58'}$$

If identical lengths $h = x_{i+1} - x_i$, $i = 1 \ldots N-1$ are chosen for the subintervals, we obtain

$$\int_a^b f(x)\,dx \approx h \cdot \left[\frac{1}{2}f(x_1) + \sum_{i=2}^{N-2} f(x_i) + \frac{1}{2}f(x_N) \right] \tag{3-58''}$$

$$h = x_{i+1} - x_i, \quad i = 1 \ldots N-1$$

Thus, we can formulate *a common rule for the discrete calculation of the coefficients of the corresponding 1D-Fourier transform* that can be used for functional approximation of measurements $[z(x_1), \ldots, z(x_N)] = [z_1, \ldots, z_N]$ in the interval $[-\pi, \pi]$. Obviously a linear combination of orthogonal trigonometric functions is used.

Let $[z(x_1), \ldots, z(x_N)] = [z_1, \ldots, z_N]$ be measurements in $[-\pi, \pi]$. These data can be approximated using the following linear combination of orthogonal trigonometric functions:

$$f_n(x) = \frac{1}{2}a_0 + \sum_{k=1}^{n} [a_k \cos(kx) + b_k \sin(kx)]$$

For the corresponding coefficients and $-\pi = x_1, x_2, n \ldots, x_N = \pi$ this yields:

$$a_0 = \frac{1}{\pi} \int_{-\pi}^{\pi} z(x)\,dx \approx \frac{1}{2\pi} \sum_{i=1}^{N-1} (z(x_i) + z(x_{i+1}))(x_{i+1} - x_i),$$

$$a_k = \frac{1}{\pi} \int_{-\pi}^{\pi} z(x)\cos(kx)\,dx \approx \frac{1}{2\pi} \sum_{i=1}^{N-1} (z(x_i)\cos(kx_i) + z(x_{i+1})\cos(kx_{i+1}))(x_{i+1} - x_i),$$

$$b_k = \frac{1}{\pi} \int_{-\pi}^{\pi} z(x)\sin(kx)\,dx \approx \frac{1}{2\pi} \sum_{i=1}^{N-1} (z(x_i)\sin(kx_i) + z(x_{i+1})\sin(kx_{i+1}))(x_{i+1} - x_i)$$

(3-59)

For a regular dissection of the interval $[-\pi, \pi]$ such as (3-58'') the equations (3-59) can be simplified to

$$a_0 = \frac{1}{\pi} \int_{-\pi}^{\pi} z(x)\,dx \approx \frac{h}{2\pi} \left[z(x_1) + 2\sum_{i=2}^{N-2} z(x_i) + z(x_N) \right],$$

$$a_k = \frac{1}{\pi} \int_{-\pi}^{\pi} z(x)\cos(kx)\,dx \approx \frac{h}{2\pi} \left[z(x_1)\cos(kx_1) + 2\sum_{i=2}^{N-2} z(x_i)\cos(kx_i) + z(x_N)\cos(kx_N) \right],$$

$$b_k = \frac{1}{\pi} \int_{-\pi}^{\pi} z(x)\sin(kx)\,dx \approx \frac{h}{2\pi} \left[z(x_1)\sin(kx_1) + 2\sum_{i=2}^{N-2} z(x_i)\sin(kx_i) + z(x_N)\sin(kx_N) \right]$$

Remark: The assumption that the measurements take values from the interval $[-\pi, \pi]$ is not an insurmountable restriction. If measurements $[z(x_1^*), \ldots, z(x_N^*)] =$

$[z_1, \ldots, z_N]$ belong to any interval $[a, b]$ we can apply the following prior transformation of the x-coordinates:

$$x_i = -\pi + \frac{x_i^* - a}{b - a} \cdot 2\pi, \; i = 1 \ldots N \tag{3-60}$$

With these new coordinates the Fourier transform for measurements $[z(x_1), \ldots, z(x_N)] = [z_1, \ldots, z_N]$ can be calculated. Finally, the back-transformation should be made in order to approximate the original measurements by a linear combination of orthogonal trigonometric functions:

$$x_i^* = a + \frac{x_i + \pi}{2\pi} \cdot (b - a), \; i = 1 \ldots N \tag{3-60'}$$

The number $2n + 1$ of the coefficients of the Fourier transform is chosen independently from the number N of given measurements.

Now we demonstrate the Fourier transform for the simple data set from Example 3.2.1.1.

Example 3.2.1.5 We consider the following temporal measurements from Example 3.2.1.1:

$$z_1 = z(1) = 0.1, \; z_2 = z(4) = 0.2, \; z_3 = z(7) = -0.1, \; z_4 = z(10) = -0.2$$

Obviously, these points do not belong to the interval $[-\pi, \pi]$, so the prior coordinate transformation (3-60) should be carried out:

$$x_i = -\pi + \frac{x_i^* - 1}{9} \cdot 2\pi, \; i = 1 \ldots 4 \tag{*.1}$$

With (*.1) the interval $[1, 10]$ is transformed into $[-\pi, \pi]$. First, we set $n = 3$ in (3-59) and calculate seven coefficients for the analytical presentation:

$$f_3(x) = \frac{1}{2}a_0 + \sum_{k=1}^{3} [a_k \cos(kx) + b_k \sin(kx)]$$

$$= \frac{1}{2}a_0 + a_1 \cos(x) + b_1 \sin(x) + a_2 \cos(2x) + b_2 \sin(2x) + a_3 \cos(3x) + b_3 \cos(3x) \tag{*.2}$$

Owing to equidistance of x we can use equation (3-59') and with $N = 4$ and $x_1 = -\pi$, $x_2 = -\frac{\pi}{3}$, $x_3 = \frac{\pi}{3}$, $x_4 = \pi$, $h = \frac{2\pi}{3}$ obtain:

$$a_0 = \frac{1}{3}[0.1 + 2 \cdot 0.2 + 2 \cdot (-0.1) + (-0.2)] = 0.0333$$

$$a_1 = \frac{1}{3}\left[0.1\cos(-\pi) + 2 \cdot 0.2 \cdot \cos\left(-\frac{\pi}{3}\right) + 2(-0.1)\cos\left(\frac{\pi}{3}\right) + (-0.2)\cos(\pi)\right] = 0.0667$$

$$b_1 = \frac{1}{3}\left[0.1\sin(-\pi) + 2 \cdot 0.2 \cdot \sin\left(-\frac{\pi}{3}\right) + 2(-0.1)\sin\left(\frac{\pi}{3}\right) + (-0.2)\sin(\pi)\right] = -0.1732$$

$$(*.3)$$

Analogously, we have: $a_2 = -0.0667$, $b_2 = -0.1732$, $a_3 = -0.0333$, $b_3 = 0$, and finally

$$f_3(x) = 0.0167 + 0.0667\cos(x) - 0.1732\sin(x)$$
$$- 0.0667\cos(2x) - 0.1732\sin(2x) - 0.0333\cos(3x) \qquad (*.4)$$

After retransformation

$$x_i^* = 1 + 9 \cdot \frac{x_i + \pi}{2\pi}, \quad i = 1 \dots 4 \qquad (*.1')$$

we obtain an analytical approximation via the Fourier transform for the given measurements (see Fig. 3.16):

$$f_3(x^*) = 0.0167 + 0.0667\cos\left(-\pi + \frac{x^*-1}{9} \cdot 2\pi\right) - 0.1732\sin\left(-\pi + \frac{x^*-1}{9} \cdot 2\pi\right)$$

$$- 0.0667\cos\left(2 \cdot \left[-\pi + \frac{x^*-1}{9} \cdot 2\pi\right]\right) - 0.1732\sin\left(2 \cdot \left[-\pi + \frac{x^*-1}{9} \cdot 2\pi\right]\right) \qquad (*.4')$$

$$- 0.0333\cos\left(3 \cdot \left[-\pi + \frac{x^*-1}{9} \cdot 2\pi\right]\right)$$

More details on the verification of the validity of the Fourier transform procedure can be found, for example, in Bracewell (1978) and Hamming (1973).

Replacing the sinusoidal and cosinusoidal functions in the orthogonal system on the interval $[-\pi, \pi]$ from Example 3.2.1.4 by other meaningful functions defined for the complete real axis that are also orthogonal and fulfill some additional important mathematical demands leads to some useful generalizations of Fourier transforms. *Wavelets* are one of these generalizations, and we now discuss the basic idea for constructing 1D-wavelets.

3.2.1.3 Approximation by 1D-Wavelets

The simplest form of a so-called *mother-wavelet* is the *hair-wavelet* Ψ_0 (see Fig. 3.17a):

$$\Psi_0(x) = \begin{cases} 1, & 0 \le x < 0.5 \\ -1, & 0.5 \le x < 1 \\ 0, & \text{sonst} \end{cases} \qquad (3\text{-}61)$$

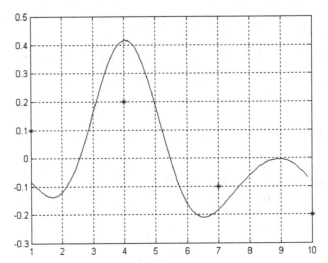

Fig. 3.16 The Fourier transform for the data from Example 3.2.1.5, $n = 3$

Any wavelet family is constructed using a certain mother-wavelet by the following transformation (see Fig. 3.17):

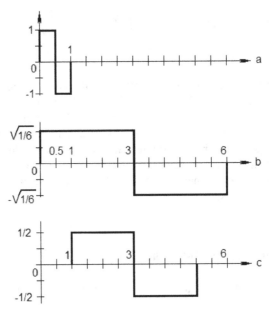

Fig. 3.17 Three hair-wavelets: (**a**) is the mother hair-wavelet $\Psi_{1,0}$ from (3-61), (**b**) is $\Psi_{6,0}$, and (**c**) is $\Psi_{4,1}$ from (3-62)

$$\Psi_{a,\,b}(x) = \frac{1}{\sqrt{a}}\Psi_0\left(\frac{x-b}{a}\right) \tag{3-62}$$

Well-known wavelet families are the Gaussian, Mexican Hat, Meyer, and Morlet families. Here we want to apply hair-wavelets to demonstrate how the wavelet approximation works. If we set

$$a = \frac{1}{2^i},\; b = \frac{k}{2^i},\; i,\,k \in Z$$

in (3-62) we obtain the orthonormal basis of the functions in $L^2(R)$:

$$\Psi_{i,\,k}(x) = \sqrt{2^i}\Psi_0\left(2^i x - k\right), \tag{3-62'}$$

Figure 3.18 shows—in a schematic rather than in a mathematically exact way— three functions from this basis. Using this basis, we can present any function $f(x)$ from the space of all twice-differentiable functions $L^2(R)$ in the following form:

$$f(x) = \sum_{i,\,k\in Z} c_{i,\,k}\Psi_{i,\,k}(x),$$
$$c_{i,\,k} = \langle f(x),\, \Psi_{i,\,k}(x)\rangle,\; i,\,k \in Z \tag{3-63}$$

After some further steps, equation (3-63) becomes

$$c_{i,\,k} = \int_{\frac{k}{2^i}}^{\frac{k+1}{2^i}} f(x)\Psi_{i,\,k}(x)\,dx,\quad i,\,k \in Z \tag{3-63'}$$

Equation (3-63') can be simplified for the case of hair-wavelets from (3-61) to

$$c_{i,\,k} = \sqrt{2^i}\int_{\frac{k}{2^i}}^{\frac{k}{2^i}+\frac{1}{2^{i+1}}} f(x)\,dx - \sqrt{2^i}\int_{\frac{k}{2^i}+\frac{1}{2^{i+1}}}^{\frac{k+1}{2^i}} f(x)\,dx,\quad i,\,k \in Z \tag{3-63''}$$

The relation from (3-63'') can be used for an approximate calculation of the weights $c_{i,\,k}$, $i,\,k \in Z$ in the case of a discrete wavelet convolution with hair-wavelets, as demonstrated in the following example.

Example 3.2.1.6 We divide the interval $[0,\,1]$ into $2^3 = 8$ subintervals

$$\left[0, \frac{1}{8}\right], \left[\frac{1}{8}, \frac{2}{8}\right] \cdots \left[\frac{7}{8}, 1\right].$$

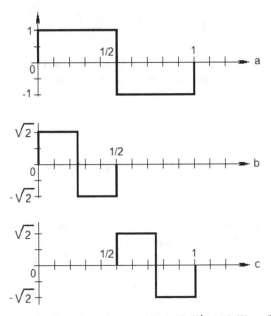

Fig. 3.18 Three hair-wavelets from the orthonormal basis (3-62′): (**a**) is $\Psi_{0,0}$, (**b**) is $\Psi_{1,0}$, and (**c**) is $\Psi_{1,1}$

The measurements are taken at the nine knots of these intervals. Thus we have:

x	0	1/8	2/8	3/8	4/8	5/8	6/8	7/8	1
z	0.0	1.707	1.0	−0.293	0.0	0.293	−1.0	−1.707	0.0

We assume that these z-values are from an unknown analytical function $f(x)$. We start with the numerical calculation of the weights using (3-63″) and the trapezium rule for the corresponding integral approximations for the interval $[0, 1]$:

$$c_{0,0} = \int_0^{1.0} f(x)\Psi_{0,0}dx = \frac{h}{2}\left[f(x_1) + 2\sum_{i=2}^{4} f(x_i) - 2f(x_5) - 2f(x_6) - 2\sum_{i=7}^{8} f(x_i) - 0 \cdot f(x_9)\right]$$

$$= \frac{h}{2}\left[z_1 + 2\sum_{i=2}^{4} z_i - 2z_5 - 2z_6 - 2\sum_{i=7}^{8} z_i\right], \quad h = \frac{1}{8}$$

$$(*.1)$$

$$c_{1,0} = \int_0^{0.5} f(x)\Psi_{1,0}dx = \int_0^{1.0} f(x)\Psi_{1,0}dx =$$

$$\frac{\sqrt{2}h}{2}[f(x_1) + 2f(x_2) - 2f(x_3) - 2f(x_4) - 2 \cdot 0 \cdot f(x_5)] = \frac{h}{\sqrt{2}}[z_1 + 2z_2 - 2z_3 - 2z_4], \quad h = \frac{1}{8}$$

$$(*.2)$$

$$c_{1,1}l = \int_{0.5}^{1} f(x)\Psi_{1,1}dx = \int_{0}^{1} f(x)\Psi_{1,1}dx$$

$$= \frac{\sqrt{2}h}{2}[2f(x_5) + 2f(x_6) - 2f(x_7) - 2f(x_8) - 0 \cdot f(x_9)] \qquad (*.3)$$

$$= \frac{h}{\sqrt{2}}[2z_5 + 2z_6 - 2z_7 - 2z_8], \quad h = \frac{1}{8}$$

$$c_{2,0} = \int_{0}^{1/4} f(x)\Psi_{2,0}dx = \int_{0}^{1} f(x)\Psi_{2,0}dx$$

$$\qquad\qquad\qquad\qquad\qquad\qquad\qquad\qquad\qquad (*.4)$$

$$= \frac{\sqrt{4}h}{2}[f(x_1) - 2f(x_2) - 2 \cdot 0 \cdot f(x_3)] = h[z_1 - 2z_2], \quad h = \frac{1}{8}$$

$$c_{2,1} = \int_{1/4}^{4/8} f(x)\Psi_{2,1}dx = \int_{0}^{1} f(x)\Psi_{2,1}dx$$

$$\qquad\qquad\qquad\qquad\qquad\qquad\qquad\qquad\qquad (*.5)$$

$$= \frac{\sqrt{4}h}{2}[2f(x_3) - 2f(x_4) - 2 \cdot 0 \cdot f(x_5)] = h[2z_3 - 2z_4], \quad h = \frac{1}{8}$$

Analogously, we obtain

$$c_{2,2} = \int_{1/2}^{6/8} f(x)\Psi_{2,2}dx = \frac{\sqrt{4}h}{2}[2f(x_5) - 2f(x_6) - 2 \cdot 0 \cdot f(x_7)] = h[2z_5 - 2z_6], \quad h = \frac{1}{8}$$

$$\qquad\qquad\qquad\qquad\qquad\qquad\qquad\qquad\qquad (*.6)$$

$$c_{2,3} = \int_{6/8}^{1} f(x)\Psi_{2,3}dx = \frac{\sqrt{4}h}{2}[2f(x_7) - 2f(x_8) - 0 \cdot f(x_9)] = h[2z_7 - 2z_8], \quad h = \frac{1}{8}$$

$$\qquad\qquad\qquad\qquad\qquad\qquad\qquad\qquad\qquad (*.7)$$

Owing to the limited number of measurements, further calculation of coefficients makes no sense. After setting the given z-values, we obtain

$$c_{0,0} = 0.60,$$
$$c_{1,0} = 0.18, \ c_{1,1} = 0.53 \qquad\qquad (*.7')$$
$$c_{2,0} = -0.43, \ c_{2,1} = 0.32, \ c_{2,2} = 0.07, \ c_{2,3} = 0.18$$

Figure 3.19 shows a discrete wavelet approximation like (3-63) for the given measurements with the obtained weights:

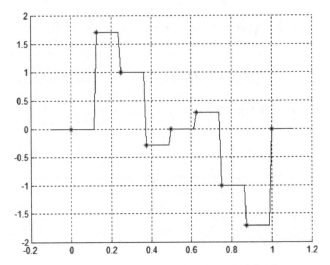

Fig. 3.19 Discrete wavelet convolution for the data from Example 3.2.1.6

$$f_3(x) = \sum_{i,\,k} \overset{c_{i,\,k}}{\Psi_{i,\,k}(x)} = c_{0,\,0}\Psi_{0,\,0}(x) + c_{1,\,0}\Psi_{1,\,0}(x) + c_{1,\,1}\Psi_{1,\,1}(x)$$

$$+ c_{2,\,0}\Psi_{2,\,0}(x) + c_{2,\,1}\Psi_{2,\,1}(x) + c_{2,\,2}\Psi_{2,\,2}(x) + c_{2,\,3}\Psi_{2,\,3}(x) \tag{*.8}$$

With the coefficients given in (*.7′), this leads to

$$f_3(x) = c_{0,\,0}[1\,(0 \le x < 0.5) - 1\,(0.5 \le x < 1)]$$
$$+ c_{1,\,0}\sqrt{2}\,[1\,(0 \le x < 1/4) - 1\,(1/4 \le x < 1/2)] + c_{1,\,1}\sqrt{2}\,[1\,(1/2 \le x < 3/4) - 1\,(3/4 \le x < 1)]$$
$$+ c_{2,\,0}2\,[1\,(0 \le x < 1/8) - 1\,(1/8 \le x < 2/8)] + c_{2,\,1}2\,[1\,(2/8 \le x < 3/8) - 1\,(3/8 \le x < 4/8)]$$
$$+ c_{2,\,2}2\,[1\,(4/8 \le x < 5/8) - 1\,(5/8 \le x < 6/8)] + c_{2,\,3}2\,[1\,(6/8 \le x < 7/8) - 1\,(7/8 \le x < 1)]$$

Certainly, there are many numerical algorithms for "optimal" wavelet convolutions: fast, quick, hard, soft, and so on. Our goal was a simple explanation of the basic idea of wavelet approximation. Further details about wavelets can be found, for example, in Strang and Nguyen (1997) and Wickerhauser (1994).

Remark: Least squares is not the only smoothing criterion that is useful for approximation approaches. Dierckx (1993) discussed an extension of the smoothing criterion for tensor product splines—the so-called variational approach—that deals with approximations where the functional form of an analytical function is not specified in advance but follows from the solution of the variational problem.

3.2.2 Stochastic Point of View: Random Processes and Useful Quantitative Characteristics

From a stochastic point of view we assume that measurements belong to a realization of a random process, which we noted in Sect. 3.1.2. A realization of a random process can be approximated by an analytical function, but a random process is a more complicated model. Often one speaks about *a random function* in regard to the yearly temperature observations or other temporal measurements, which are interpreted as random process. The realization (or trajectory) of this process is a curve for each particular year. A value of this random process on a certain day is a random variable. Definition 3.1.2-1 for random fields can be adapted for random functions follows:

Definition 3.2.2-1 A family of random variables $Z(t) = \{Z_t,\ t \in T\}$ is called a *random process* or a *random function*. With $T = \{0,\ 1,\ 2,\ \ldots\}$ this random process is called *discrete*. With $T = [0,\ \infty)$ it is designated as *continuous*.

Definition 3.2.2-1 leads to random processes. As an additional characteristic the *k-dimensional distribution function* can be defined for random processes:

Definition 3.2.2-2 For any choice of $t_1, \ldots,\ t_k \in T$ the general distribution $F(x_1, \ldots,\ x_k)$
$= P(Z(t_1) < x_1, \ldots,\ Z(t_k) < x_k)$ of a random vector is called the *k-dimensional distribution function*.

Remark: Random processes with *k*-dimensional normal distribution are called *Gaussian processes*.

Let us discuss some important models of weakly stationary (here, in short, stationary) random processes. We speak about *weakly stationary random processes* if the following two demands are fulfilled:

$$E(Z(t)) = \mu = const,$$
$$Cov(Z(t_i),\ Z(t_j)) = E(Z(t_i) - \mu)(Z(t_j) - \mu) = c(t_j - t_i) = c(\tau), \qquad (3\text{-}64)$$
$$Var(Z(t)) = c(0) = \sigma^2, \quad t_i,\ t_j,\ \tau \in T$$

These conditions lead to a process with a constant mean and covariance function that depend solely on the distance between the two random values from this process.

The "Pure" Random Process (White-Noise-Process)

This discrete model describes the case of total stochastic independence of identically distributed measurements. The random values $\varepsilon_i = Z(t_i)$, $t_i \in \{0,\ 1,\ 2 \ldots\}$ of this process can be denoted as *i.i.d.* (*independently identically distributed*) They follow an identical distribution function $F_0(x)$. Thus, for this case the following holds:

$$\forall k: \quad F(x_1, \dots, x_k) = F_0(x_1) \cdot \dots \cdot F_0(x_k)$$
$$c(\tau) = \begin{cases} \sigma^2, & \tau = 0 \\ 0, & \tau \neq 0 \end{cases} \tag{3-65}$$

Figure 3.20a shows a realization of this process. Here we use the normal distribution with $\mu = 0$, $\sigma^2 = 1$ for measurements at locations from $T = \{0, 1, 2, \dots\}$.

Remark: A possible simple—nonstationary—generalization of this process can be constructed if the demand of identical distribution is omitted. For example, we have measurements where the mean and the variance vary depending on t. Thus, we have

$$c(\tau, t) = \begin{cases} \sigma^2(t), & \tau = 0 \\ 0, & \tau \neq 0 \end{cases} \tag{3-65'}$$

Figure 3.20b shows a realization of this process. Here we use normal distribution with $\mu = 0$, $\sigma^2 = \exp(-0.05t)$ for measurements at locations from $T = \{0, 1, 2, \dots\}$.

Moving-Average or MA(1) Process

In this case, the random value at location t is the following weighted sum of neighboring random values of the white-noise process (3-65):

$$Z(t) = \varepsilon(t) + a \cdot \varepsilon(t-1) \tag{3-66}$$

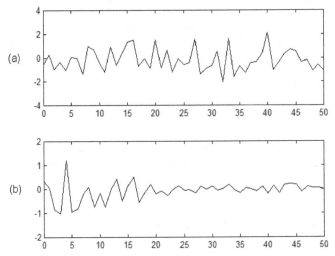

Fig. 3.20 Realizations of a stationary white-noise process (**a**), and a nonstationary white-noise process (**b**). Normal distribution is used with $\mu = 0$, $\sigma^2 = 1$ in (**a**) and $\mu = 0$, $\sigma^2 = \exp(-0.05t)$ in (**b**)

The parameter a is a deterministic constant. If we assume $E\left(\varepsilon\left(t\right)\right)=0$, $Var\left(\varepsilon\left(t\right)\right)=\sigma^2$, we have

$$c_Z\left(\tau\right)=\begin{cases}\left(1+a^2\right)\sigma^2, & \tau=0\\ a\sigma^2, & \tau=\pm1\\ 0, & \tau=\pm2,\ \pm3,\ldots\end{cases}\tag{3-67}$$

MA(q) Process

A logical generalization of the MA(1) process (3-66) is the MA(q) process taking into account k neighboring values of the white-noise process. This is obviously expressed in the following way:

$$Z\left(t\right)=\varepsilon\left(t\right)+a_1\cdot\varepsilon\left(t-1\right)+\ldots+a_q\varepsilon\left(t-q\right)\tag{3-66'}$$

Autoregressive Process AR(1)

In AR(1) processes the random value at t is the weighted sum of the prior random value of the process, that is, $Z\left(t-1\right)$, and the value of the white-noise process $\varepsilon\left(t\right)$. This means that

$$Z\left(t\right)=aZ\left(t-1\right)+\varepsilon\left(t\right)\tag{3-68}$$

AR(1) is stationary only if the deterministic constant fulfills $|a|<1$. This process is a random model of simple linear regression. It is also true that

$$E\left(Z\left(t\right)\right)=0,\ Var\left(Z\left(t\right)\right)=\frac{\sigma^2}{1-a^2},$$
$$c_Z\left(\tau\right)=a^{|\tau|}\frac{\sigma^2}{1-a^2}\tag{3-69}$$

Figure 3.21 shows two realizations of the same random process AR(1). The initial random variable $Z\left(0\right)$ is normally distributed with $\mu=5$, $\sigma^2=1$. For the white-noise process $\mu=0$, $\sigma^2=0.5$ are chosen. We use $a=0.8$ as the constant.

ARMA(p, q) Process

An ARMA(p, q) process generalizes AR(1) as well as MA(q)-processes. When both approaches are taken into account, this generalization is given by

$$Z\left(t\right)=a_1Z\left(t-1\right)+\ldots+a_pZ\left(t-p\right)+\varepsilon\left(t\right)+b_1\varepsilon\left(t-1\right)+\ldots+b_q\varepsilon\left(t-q\right)\tag{3-70}$$

With constants fulfilling some demands this process is stationary.

But how can temporal measurements really be analyzed? We discuss some approaches from a special statistical framework known as *time series analysis*. The term "time series" describes a temporal ordered sequence of quantitative

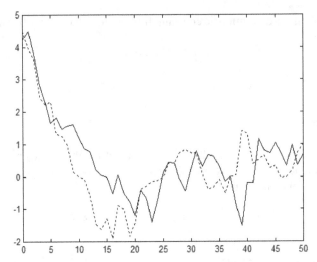

Fig. 3.21 Two realizations of the same random process AR(1). The initial random variable $Z(0)$ is Gaussian with $\mu = 5$, $\sigma^2 = 1$. For the applied white-noise process $\mu = 0$, $\sigma^2 = 0.5$ are chosen. We use $a = 0.8$ as the constant

measurements. Practical examples for time series might include air temperatures, rainfall volumes, wind forces, and stock prices.

First, we assume that given measurements can be described by the following simple additive model:

$$Z(t) = m(t) + \varepsilon(t) \tag{3-71}$$

where $m(t)$ denotes a deterministic and unknown trend. The random component $\varepsilon(t)$ describes the residuals. We can estimate this unknown trend with the simple linear form $m(t) = a_0 + a_1 t$ The unknown parameters a_0 and a_1 can be estimated by applying the least-squares method, which leads to $\hat{m}(t) = \hat{a}_0 + \hat{a}_1 t$. Estimating these parameters follows the same course as in linear regression approaches. Therefore, for N measurements, $(t_1, Z(t_1)), \ldots, (t_N, Z(t_N))$ yields the following estimates:

$$\hat{a}_1 = \frac{\sum\limits_{i=1}^{N} (t_i - \bar{t})(Z(t_i) - \bar{m})}{\sum\limits_{i=1}^{N} (Z(t_i) - \bar{m})^2}, \quad \bar{t} = \frac{\sum\limits_{i=1}^{N} t_i}{N}, \quad \bar{m} = \frac{\sum\limits_{i=1}^{N} Z(t_i)}{N}, \tag{3-72}$$

$$\hat{a}_0 = \bar{t} - \hat{a}_1 \bar{m}$$

In a more generalized approach any unknown trend can be described as a so-called quasi-linear model $m(t) = a_0 + a_1 m_1(t) + \ldots + a_d m_d(t)$, which can be applied with any time-dependent function $m_1(t), \ldots, m_d(t)$ with unknown parameters $a_0, a_1, \ldots, a_d, d \leq N - 1$. Obviously, these parameters have to be estimated using the least-squares method. Remember that this is the kind of deterministic problem that was discussed in Sect. 3.2.1. After finishing the estimation process leading to $\hat{a}_0$, $\hat{a}_1$, ..., $\hat{a}_d$ and the corresponding trend $\hat{m}(t) = \hat{a}_0 + \hat{a}_1 m_1(t) + \ldots + \hat{a}_d m_d(t)$, we can prove the "quality" or goodness of fit of the chosen model by considering the following characteristics:

$$B = 1 - \frac{\sum\limits_{i=1}^{N}(Z(t_i) - \hat{m}(t_i))^2}{\sum\limits_{i=1}^{N}(Z(t_i) - \bar{m})^2}, \qquad \hat{s}^2 = \frac{1}{N-2}\sum\limits_{i=1}^{N}(Z(t_i) - \hat{m}(t_i))^2 \qquad (3\text{-}73)$$

If parameter B is nearly one, the linear (quasi-linear) model can be accepted. From a statistical point of view B is called the *coefficient of determination*. The second characteristic $\hat{s}^2$ is an estimator for the unknown variance of the residuals.

The residuals $\hat{\varepsilon}(t) = Z(t) - \hat{m}(t) = Z(t) - \hat{a}_0 - \hat{a}_1 m_1(t) - \ldots - \hat{a}_d m_d(t)$ should also be analyzed, and we now discuss this topic.

Example 3.2.2.1 We consider the same measurements as in Example 3.2.1.6:

t	0	1/8	2/8	3/8	4/8	5/8	6/8	7/8	1
$Z(t)$	0.0	1.707	1.0	−0.293	0.0	0.293	−1.0	−1.707	0.0

At first, we assume that these measurements follow the additive model (3-71) and the unknown trend, fulfilling

$$m(t) = a_0 + a_1 \sin(2\pi t) \qquad (*.1)$$

We apply the least-squares method for estimating a_0, a_1 as in Example 3.2.1.1:

$$\begin{aligned} F(a_0, a_1) &= \sum\limits_{i=1}^{9}(a_0 + a_1 \sin(2\pi t_i) - Z(t_i))^2 \\ &= (a_0 + a_1 \sin(2\pi \cdot 0) - 0.0)^2 + \cdots \\ &\quad + (a_0 + a_1 \sin(2\pi \cdot 1) - 0.0)^2 \rightarrow min(a_0, a_1) \end{aligned} \qquad (*.2)$$

Regarding partial derivatives with respect to a_0 and a_1 and setting these derivatives equal to zero leads to a LSE where the solution of the corresponding minimizing problem is given at $\hat{a}_0$, $\hat{a}_1$:

$$\begin{pmatrix} 9 & \sum\limits_{i=1}^{9}\sin(2\pi t_i) \\ \sum\limits_{i=1}^{9}\sin(2\pi t_i) & \sum\limits_{i=1}^{9}\sin^2(2\pi t_i) \end{pmatrix} \begin{pmatrix} \hat{a}_0 \\ \hat{a}_1 \end{pmatrix} = \begin{pmatrix} \sum\limits_{i=1}^{9} Z(t_i) \\ \sum\limits_{i=1}^{9} Z(t_i)\sin(2\pi t_i) \end{pmatrix} \qquad (*.3)$$

Using the given values and solving the LSE (*.3) leads to

$$\begin{pmatrix} 9 & 0 \\ 0 & 4 \end{pmatrix} \begin{pmatrix} \hat{a}_0 \\ \hat{a}_1 \end{pmatrix} = \begin{pmatrix} 0 \\ 3.9997 \end{pmatrix} \Rightarrow \begin{pmatrix} \hat{a}_0 \\ \hat{a}_1 \end{pmatrix} = \begin{pmatrix} 9 & 0 \\ 0 & 4 \end{pmatrix}^{-1} \begin{pmatrix} 0 \\ 3.9997 \end{pmatrix} \qquad (*.3')$$

and $\hat{a}_0 = 0.0$, $\hat{a}_1 = 0.9999$. Thus, the unknown trend can be estimated by $\hat{m}(t) = 0.9999 \sin(2\pi t)$.

Now, should we accept this quasi-linear model or not? In order to answer this question we should calculate the parameter B from (3-73) and assess it compared with one:

$$B = 1 - \frac{\sum\limits_{i=1}^{9} (Z(t_i) - 0.9999 \sin(2\pi t_i))^2}{\sum\limits_{i=1}^{9} (Z(t_i) - \bar{m})^2} = 1 - \frac{4}{7.9994} = 0.5, \quad \bar{m} = \frac{1}{9}\sum\limits_{i=1}^{9} Z(t_i) = 0$$

$$(*.4)$$

Obviously, the parameter B that we obtain is not approximately equal to one. The quasi-linear model (*.1) does not provide a perfect or even a sufficient fit, so let us try to improve upon it by assuming the following approach:

$$m(t) = a_0 + a_1 \sin(2\pi t) + a_2 \sin(4\pi t) \qquad (*.5)$$

Once again, we have to apply the least-squares method to estimate of a_0, a_1, a_2 following the common principle from Sect. 3.2.1:

$$F(a_0, a_1, a_2) = \sum\limits_{i=1}^{9} (a_0 + a_1 \sin(2\pi t_i) + a_2 \sin(4\pi t_i) - Z(t_i))^2$$

$$= (a_0 + a_1 \sin(2\pi \cdot 0) + a_2 \sin(4\pi \cdot 0) - 0.0)^2 + \dots$$

$$+ (a_0 + a_1 \sin(2\pi \cdot 1) + a_2 \sin(4\pi \cdot 1) - 0.0)^2 \rightarrow min(a_0, a_1, a_2)$$

$$(*.6)$$

Partially differentiating the function in (*.6) with respect to a_0, a_1, a_2 in order to solve the corresponding minimization problem leads to the following LSE:

$$\begin{pmatrix} 9 & \sum\limits_{i=1}^{9} \sin(2\pi t_i) & \sum\limits_{i=1}^{9} \sin(4\pi t_i) \\ \sum\limits_{i=1}^{9} \sin(2\pi t_i) & \sum\limits_{i=1}^{9} \sin^2(2\pi t_i) & \sum\limits_{i=1}^{9} \sin(2\pi t_i)\sin(4\pi t_i) \\ \sum\limits_{i=1}^{9} \sin(4\pi t_i) & \sum\limits_{i=1}^{9} \sin(2\pi t_i)\sin(4\pi t_i) & \sum\limits_{i=1}^{9} \sin^2(4\pi t_i) \end{pmatrix} \begin{pmatrix} \hat{a}_0 \\ \hat{a}_1 \\ \hat{a}_2 \end{pmatrix} = \begin{pmatrix} \sum\limits_{i=1}^{9} Z(t_i) \\ \sum\limits_{i=1}^{9} Z(t_i)\sin(2\pi t_i) \\ \sum\limits_{i=1}^{9} Z(t_i)\sin(4\pi t_i) \end{pmatrix}$$

$$(*.7)$$

Using the given values and solving LSE (*.7) yields

$$\begin{pmatrix} 9 & 0 & 0 \\ 0 & 4 & 0 \\ 0 & 0 & 4 \end{pmatrix} \begin{pmatrix} \hat{a}_0 \\ \hat{a}_1 \\ \hat{a}_2 \end{pmatrix} = \begin{pmatrix} 0 \\ 3.9997 \\ 4 \end{pmatrix} \Rightarrow \begin{pmatrix} \hat{a}_0 \\ \hat{a}_1 \\ \hat{a}_2 \end{pmatrix} = \begin{pmatrix} 9 & 0 & 0 \\ 0 & 4 & 0 \\ 0 & 0 & 4 \end{pmatrix}^{-1} \begin{pmatrix} 0 \\ 3.9997 \\ 4 \end{pmatrix} = \begin{pmatrix} 0 \\ 0.9999 \\ 1.0000 \end{pmatrix}$$

$$(*.7')$$

and $\hat{a}_0 = 0.0$, $\hat{a}_1 = 0.9999$, $\hat{a}_2 = 1.0$.

The unknown trend can be estimated by $\hat{m}(t) = 0.9999 \sin(2\pi t) + \sin(4\pi t)$. Can this improved quasi-linear model be accepted? In order to answer this question we should again calculate the coefficient of determination B from (3-73) and compare it with the optimal value equal to one:

$$B = 1 - \frac{\sum\limits_{i=1}^{9} (Z(t_i) - 0.9999\sin(2\pi t_i) - \sin(4\pi t_i))^2}{\sum\limits_{i=1}^{9} (Z(t_i) - \bar{m})^2} = 1 - \frac{2.28 \cdot 10^{-8}}{7.9994} \approx 1,$$

$$\bar{m} = \frac{1}{9}\sum_{i=1}^{9} Z(t_i) = 0$$

(*.8)

Parameter B is now approximately equal to one so the quasi-linear model (*.5) is better than (*.1). Figure 3.22 shows the estimated trend for quasi-linear models (*.1) and (*.5).

Remark: Wavelet approaches and Fourier transforms can be also applied for trend fitting (trend estimation); see Sect. 3.2.1 for further details. More about stochastic processes can be found in Chiang (1980).

There are many different approaches to ensure trend fitting, trend smoothing, and trend elimination. We discuss some of these related to their fields of application.

3.2.2.1 Moving Average

The moving average method, which is based on a locally adapted construction of arithmetical means, is used for smoothing and filtering time series. We start with

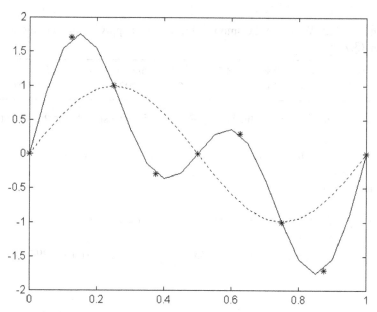

Fig. 3.22 Estimated trend for quasi-linear models (*.1, *dashed line*) and (*.5, *solid line*) for the data set (*stars*) from Example 3.2.2.1

a time series without a so-called seasonal component, which means that a periodic function or seasonal component $s(t)$ vanishes in the generalized presentation of (3-71):

$$Z(t) = m(t) + \varepsilon(t) = g(t) + s(t) + \varepsilon(t),$$
$$\exists p: \quad s(t + pk) = s(t), \ k = \pm 1, \pm 2, \ldots \qquad (3\text{-}71')$$

A mathematical model of time series without seasonal components corresponds to

$$Z(t) = g(t) + \varepsilon(t) \qquad (3\text{-}74)$$

Now, we formulate the *common rule of the moving average of the $(2m+1)$ order for time series without a seasonal component*:

Let $(t_1, Z(t_1)), \ldots, (t_N, Z(t_N))$ be measurements belonging to a time series. For a given parameter m: $m < (N-1)/2$ we call the new time series given by

$$\bar{Z}(t_i) = \frac{1}{2m+1} \sum_{j=-m}^{m} Z(t_{i+j}), \ i \geq m+1, \ \ldots, N-m \qquad (3.75)$$

the *moving average* of the $(2m+1)$ order.

We prove how rule (3.75) works using the data set from Example 3.2.2.1 for $m = 1$ and $m = 2$.

Example 3.2.2.2 We obtain the smoothed time series applying the moving average method (3.75):

t	0	1/8	2/8	3/8	4/8	5/8	6/8	7/8	1
$Z(t)$	0.0	1.707	1.0	−0.293	0.0	0.293	−1.0	−1.707	0.0

• $m = 1$: Moving average of third order. Rule (3.75) corresponds for this case to

$$\bar{Z}(t_i) = \frac{1}{3}[Z(t_{i-1}) + Z(t_i) + Z(t_{i+1})], \ i \geq 2, \ldots 8 \qquad (*.1)$$

For example, we get

$$\bar{Z}(1/8) = \frac{1}{3}[Z(0) + Z(1/8) + Z(2/8)] = \frac{1}{3}[0.0 + 1.707 + 1.0] = 0.902,$$

$$\bar{Z}(4/8) = \frac{1}{3}[Z(3/8) + Z(4/8) + Z(5/8)] = \frac{1}{3}[-0.293 + 0.0 + 0.293] = 0.0$$

The smoothed time series is given by

t	1/8	2/8	3/8	4/8	5/8	6/8	7/8
$\bar{Z}(t)$	0.902	0.805	0.236	0.0	−0.236	−0.805	−0.902

as shown in Fig. 3.23a.

• $m = 2$: Moving average of fifth order. Rule (3.75) leads to

$$\bar{Z}(t_i) = \frac{1}{5}[Z(t_{i-1}) + Z(t_i) + Z(t_{i+1})], \; i \geq 3, \dots 7 \qquad (*.2)$$

For example,

$$\bar{Z}(2/8) = \frac{1}{5}[Z(0) + Z(1/8) + Z(2/8) + Z(3/8) + Z(4/8)] = 0.483,$$

$$\bar{Z}(4/8) = \frac{1}{3}[Z(2/8) + Z(3/8) + Z(4/8) + Z(5/8) + Z(6/8)] = 0.0$$

holds. Smoothed time series are shorter and less rough than the original ones.
    The smoothed time series corresponds to

t	2/8	3/8	4/8	5/8	6/8
$\bar{Z}(t)$	0.483	0.541	0.0	−0.541	−0.483

and is shown in Fig. 3.23b, where it can be seen that smoothing time series applying so-called *exponential or geometrical smoothing* plays an important role, especially for predicting future values of the time series. This method is based on weighted arithmetical means.

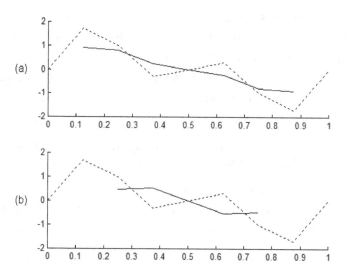

Fig. 3.23 Smoothed time series (solid line) by the moving average of the third order (**a**) and of fifth order (**b**). Original time series are drawn as dashed lines

3.2.2.2 Geometrical Smoothing of Time Series

We start with a common rule:

Let $Z(t)$ describe a time series. For a given parameter $\alpha : 0 < \alpha < 1$ we call the new time series generated by

$$\bar{Z}(t_i) = \sum_{j=0}^{\infty} \alpha (1 - \alpha)^j Z(t_{i-j}), \quad j = 0, 1, \ldots \qquad (3\text{-}76)$$

geometrical smoothing with smoothing parameter α. It is advisable to choose $\alpha : 0.2 < \alpha < 0.3$. Based on measurements $(t_1, Z(t_1)), \ldots, (t_N, Z(t_N))$ the one-step prediction can be obtained by

$$\hat{Z}(t_{N+1}) = \alpha Z(t_N) + \alpha (1 - \alpha) Z(t_{N-1}) + \ldots + \alpha (1 - \alpha)^{N-1} Z(t_1) \quad (3\text{-}76')$$

The following example presents an application of geometrical smoothing ensuring smoothing and one-step prediction.

Example 3.2.2.3 The original data remain the same:

t	0	1/8	2/8	3/8	4/8	5/8	6/8	7/8	1
$Z(t)$	0.0	1.707	1.0	−0.293	0.0	0.293	−1.0	−1.707	0.0

We choose the smoothing parameter $\alpha = 0.25$. Recursively using (3-76) leads to

$$\bar{Z}(0) = \alpha Z(0) = 0.25 \cdot Z(0) = 0$$
$$\bar{Z}(1/8) = \alpha Z(1/8) + \alpha (1 - \alpha) Z(0) = 0.25 \cdot Z(1/8) + 0.25 (1 - 0.25) \cdot Z(0) = 0.427$$
$$\bar{Z}(2/8) = \alpha Z(2/8) + \alpha (1 - \alpha) Z(1/8) + \alpha (1 - \alpha)^2 Z(0)$$
$$= 0.25 \cdot Z(2/8) + 0.25 (1 - 0.25) \cdot Z(1/8) + 0.25 (1 - 0.25)^2 Z(0) = 0.5701$$
$$\bar{Z}(1) = \alpha Z(1) + \alpha (1 - \alpha) Z(7/8) + \alpha (1 - \alpha)^2 Z(6/8) + \ldots + \alpha (1 - \alpha)^8 Z(0)$$
$$= 0.25 \cdot Z(1) + 0.25 (1 - 0.25) \cdot Z(7/8) + \ldots + 0.25 (1 - 0.25)^8 Z(0) = -0.346$$

Figure 3.24 shows this geometrical smoothing. A one-step prediction at point $t = 1 + 1/8 = 1.125$ can be achieved using (3-76'):

$$\hat{Z}(1.125) = \alpha Z(1) + \alpha (1 - \alpha) Z(7/8) + \ldots + \alpha (1 - \alpha)^8 Z(0)$$
$$= 0.25 \cdot Z(1) + 0.25 (1 - 0.25) \cdot Z(7/8) + \ldots + 0.25 (1 - 0.25)^8 \cdot Z(0)$$
$$\times = \bar{Z}(1) = -0.346$$

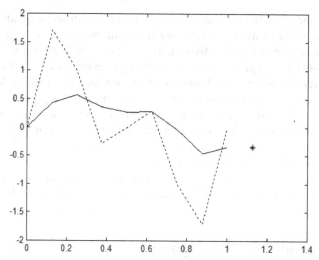

Fig. 3.24 Geometrical smoothing of the time series (*solid line*) and one-step prediction at point $t = 1.125$ (*marked as star*). The original measurements are given by the dashed line

3.2.2.3 Consideration of a Seasonal Component

If the period p of a seasonal component in (3-71') is known, the influence of this component can be eliminated by using a moving average with a specially adapted order m. For simplification, let us assume that $t = 0, 1, 2, \ldots$ and $p = 4$ (or another even p, that is, $p = 2k, k = 0, 1, 2 \ldots$) for a time series $Z(t)$. After smoothing $Z(t)$ by the moving average method (3.75) with $m = p/2$ and *specially weighted* end-values, we obtain:

$$\bar{Z}(t_i) = \frac{1}{p} \left[\frac{1}{2} Z(t-m) + Z(t-m+1) + \ldots + Z(t) + \ldots + Z(t+m-1) + \frac{1}{2} Z(t+m) \right]$$

$$= \frac{1}{4} \left[\frac{1}{2} Z(t-2) + Z(t-1) + .Z(t) + Z(t+1) + \frac{1}{2} Z(t+2) \right]$$

If $p = 2k + 1$ for $k = 0, 1, 2 \ldots$, then $m = k$ can be used in (3.75).

Remark: There is a so-called *variate difference method* for polynomial trend elimination, which leads to special asymmetrical filters. Some alternative methods such as phase mean and Fourier transform can be used for elimination or estimation of seasonal components with known period p. More details about time series analysis can be found in Box and Jenkins (1976) and Brockwell and Davis (1991).

3.2.2.4 Time Series Modeling Using Stochastic Processes

We have already defined a stochastic process, especially a stationary stochastic process. We now want to discuss some well-known models of stationary stochastic processes. Time series with a seasonal component or with different trends should be considered as realizations of nonstationary stochastic processes. For example, there are so-called nonstationary ARIMA(p, d, q) processes, which are generalizations of stationary ARMA(p, q) processes. Each modeling or even describing of time series starts with trend elimination and transfer of the nonstationary case to the stationary. Some smoothing approaches that help to treat the trend $m(t)$ in a time series $Z(t) = m(t) + \varepsilon(t)$ have already been explained. Now we discuss a basic idea for analyzing residuals $\varepsilon(t)$, which represent the random or stochastic part of the time series. We assume that these residuals belong to a finite realization of a stationary stochastic process $Z_0(t)$ with discrete time $t \in T$.

In order to estimate characteristics of a stationary process based on a single realization of residuals an assumption about the so-called *ergodicity* of the process is necessary:

$$\lim_{N \to \infty} E\left(\bar{Z}_0 - \mu\right)^2 = 0, \quad \bar{Z}_0 = \frac{1}{N} \sum_{i=1}^{N} Z_0(t_i)$$

$$\lim_{N \to \infty} E\left(C_k^* - c(t_i, t_{i+k})\right)^2 = 0 \quad C_k^* = \frac{1}{N} \sum_{i=1}^{N} (Z_0(t_i) - \mu)(Z_0(t_{i+k}) - \mu)$$

(3-77)

This assumption can be considered as justification for working with a single realization. It is assumed that this realization is "usual" with respect to the stochastic process and does not represent an outlier. Using this fact, we can estimate the mean and covariance function of the stochastic process $Z_0(t)$ by

$$\bar{Z}_0 = \frac{1}{N} \sum_{i=1}^{N} Z_0(t_i)$$

$$\hat{c}(t_i, t_{i+k}) = \hat{c}_k = \frac{1}{N} \sum_{i=1}^{N-k} (Z_0(t_i) - \bar{Z}_0)(Z_0(t_{i+k}) - \bar{Z}_0) = \hat{c}_{-k}$$

(3-78)

These characteristics are called the *empirical mean* and *empirical covariance*. The empirical variance and empirical correlation function can be calculated by

$$\hat{c}_0 = \frac{1}{N} \sum_{i=1}^{N} (Z_0(t_i) - \bar{Z}_0)^2$$

$$\hat{\rho}(t_i, t_{i+k}) = \hat{\rho}_k = \frac{\hat{c}_k}{\hat{c}_0} = \hat{\rho}_{-k}$$

(3-79)

It should be noted that empirical covariance and the empirical correlation function should be determined for $k < N/4$ because otherwise a sufficient number of pairs is not taken into account. A presentation of the empirical correlation function $\hat{\rho}_k$, $k = 1, 2, \ldots$ is called a *correlogram*. Comparison with some known correlogram models of special stationary processes allows us to make a first decision on a possible and useful model. Generally statistical software disposes of different tools for model fitting.

We illustrate some important steps of modeling and calculating (3-78) and (3-79) for a simple data set with which we are already familiar.

Example 3.2.2.4 Let us consider the following short time series:

t	0	1/8	2/8	3/8	4/8	5/8	6/8	7/8	1
$Z(t)$	0.0	1.707	1.0	−0.293	0.0	0.293	−1.0	−1.707	0.0

We interpret this realization as one of a nonstationary process

$$Z(t) = m(t) + Z_0(t) \tag{*.1}$$

and we want to identify the kind of stochastic process with which we are dealing. The first step in modeling—fitting the deterministic structure or the trend $m(t)$—was done in Example 3.2.2.1, and the following quasi-linear trend model was assumed, estimated, and tested:

$$\hat{m}(t) = 0.9999 \sin(2\pi t) + \sin(4\pi t) \tag{*.2}$$

This trend approximation can be accepted because of parameter $B \approx 1$. Now we begin the second step. We consider the residuals, which we have calculated easily using the obvious relation

$$\varepsilon(t_i) = Z(t_i) - \hat{m}(t_i), \, i = 1, \ldots, 9 \tag{*.3}$$

and we obtain:

t	0	1/8	2/8	3/8	4/8	5/8	6/8	7/8	1
$\varepsilon(t) \cdot 10^3$	0.0	−0.036	0.1	−0.036	0.0	0.036	−0.1	0.036	0.0

Thus, in this step we assume that these residuals are a finite realization of a stationary stochastic process $Z_0(t)$ and we begin to estimate the model parameters. With (3-78) the unknown mean of this stationary process can be estimated by

$$\bar{Z}_0 = \frac{1}{9} \sum_{i=1}^{9} Z_0(t_i) = \frac{1}{9} \sum_{i=1}^{9} \varepsilon(t_i) = 0.0 \tag{*.4}$$

We use equations (3-78) and (3-79) and take the fact that $k < 9/4 \Rightarrow k = 1, 2$ into account. This leads to

$$\hat{c}_0 = \frac{1}{9} \sum_{i=1}^{9} (Z_0(t_i) - \bar{Z}_0)^2 = \frac{1}{9} \sum_{i=1}^{9} (\varepsilon(t_i) - \bar{Z}_0)^2 = \frac{1}{9} \sum_{i=1}^{9} \varepsilon^2(t_i) = 2.8 \cdot 10^{-9} \tag{*.5}$$

There are eight pairs that can be considered for calculating $\hat{c}_1$:

$$\hat{c}_1 = \frac{1}{9} \sum_{i=1}^{9-1} (Z_0(t_i) - \bar{Z}_0)(Z_0(t_{i+1}) - \bar{Z}_0) = \frac{1}{8} \sum_{i=1}^{8} (\varepsilon(t_i) - 0)(\varepsilon(t_{i+1}) - 0)$$

$$= \frac{1}{9} \sum_{i=1}^{8} \varepsilon(t_i)\varepsilon(t_{i+1}) = \frac{10^{-6}}{9} [0.0 \cdot (-0.036) + (-0.036) \cdot 0.1 + \dots + 0.036 \cdot 0.0] = -1.8 \cdot 10^{-9}$$

$$(*.6)$$

The estimation $\hat{c}_2$ of the covariance for $k = 2$ is based on seven pairs:

$$\hat{c}_2 = \frac{1}{9} \sum_{i=1}^{9-2} (Z_0(t_i) - \bar{Z}_0)(Z_0(t_{i+2}) - \bar{Z}_0) = \frac{1}{7} \sum_{i=1}^{7} (\varepsilon(t_i) - 0)(\varepsilon(t_{i+2}) - 0)$$

$$= \frac{1}{9} \sum_{i=1}^{7} \varepsilon(t_i)\varepsilon(t_{i+2}) = \frac{10^{-6}}{9} [0.0 \cdot 0.1 + (-0.036)^2 + \dots + (-0.1) \cdot 0.0] = 1.9 \cdot 10^{-10}$$

$$(*.6')$$

Thus, the empirical correlation function for $k = 0, 1, 2$ is

$$\hat{\rho}_k = \frac{\hat{c}_k}{\hat{c}_0}, \quad k = 0, 1, 2 \Rightarrow$$

$$\hat{\rho}_0 = 1, \ \hat{\rho}_1 = -0.57, \ \hat{\rho}_2 = 0.05$$

$$(*.7)$$

Figure 3.25 shows the empirical correlation function.

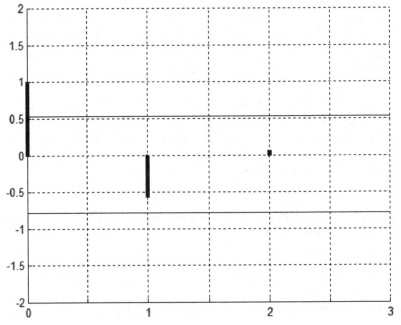

Fig. 3.25 Empirical correlation function (*thick solid line*) calculated in Example 3.2.2.4 with a confidence band given by a lower line g_1 and an upper line g_2

In the third step, model fitting, the model should be identified by comparison with well-known theoretical models of correlation functions. Unfortunately, in our case the time series is too short to allow us to finish our modeling. But this example is continued below.

Remark: Model fitting can be completed using special test approaches. For example, one can test whether the stationary process chosen for residuals is a white-noise process or not. For this, we make use of the fact that for a white-noise process the values of the empirical correlation function are independently and identically distributed following the normal distribution with $\hat{\rho}_k \sim N\left(\frac{1}{N}, \frac{1}{N}\right)$

The limit values for the confidence interval—or significance band—for each $k = 1, 2$ using a significant value α can be given (see Fig. 3.25):

$$g_1 = -\frac{1}{N} - \frac{u_{1-\alpha/2}}{\sqrt{N}}, \quad g_2 = -\frac{1}{N} + \frac{u_{1-\alpha/2}}{\sqrt{N}}$$

$$\Phi\left(u_{1-\alpha/2}\right) = 1 - \alpha/2, \quad \Phi(x) = \frac{1}{\sqrt{2\pi}} \int_{-\infty}^{x} \exp\left(-\frac{t^2}{2}\right) dt$$

The hypothesis about a white-noise process is not rejected if

$$g_1 \leq \hat{\rho}_k \leq g_2, \quad \forall k = 1, 2, \ldots$$

3.2.2.5 Forecasting with a Fitted Model

After model identification, estimation of model parameters, and model fitting we can predict an unknown, future value $Z(t_i), i > N$ of the time series $Z(t_i), i = 1, 2, \ldots, N$. This means that we can make a forecast based on the fitted model. For this we use the fact—confirmed by tests—that our time series follows the "rule"

$$Z(t) = \hat{m}(t) + Z_0(t) \tag{3-80}$$

Here the trend is estimated and a stochastic process $Z_0(t)$ is fitted by a well-known stationary one. Generally, statistical tools offer many such process models, and forecasting is done by simulating new values of $Z(t_i), i > N$.

3.2.2.6 Continuation of Example 3.2.2.4

We assume that the hypothesis about a white-noise process was not rejected. We cannot prove it using such a short time series, but it is really true because this time series from Example 3.2.2.4 is generated by the author by this way: It is a realisation of a white-noise process. Equation (3-80) corresponds to the following [cf. (*.5)] from Example 3.2.2.4:

$$Z(t) = 0.9999 \sin(2\pi t) + \sin(4\pi t) + Z_0(t),$$
$$Z_0(t) \sim N(0, \hat{c}_0) = N(0, 2.8 \cdot 10^{-9})$$
(*.8)

The stationary stochastic process $Z_0(t)$ is a white-noise process with independently, identically, and normally distributed random values for each t. The forecasting for "future" t from an interval $[1, 2]$ can be obtained following (*.8). The result is given in Fig. 3.26a.

Owing to the very small variance of the stationary stochastic process $Z_0(t)$, its influence on this forecasting cannot really be seen in Fig. 3.26a. Let us underscore the drastically different orders of the deterministic trend values and the values coming from the stationary stochastic process $Z_0(t)$. We show two realizations of $Z_0(t)$ that are added to this trend by forecasting separately in Fig. 3.26b.

It is interesting to have a look at the course of another model for the same time series. Let us consider the—not optimal but possible—model of stochastic processes $Z(t) = \hat{m}(t) + Z_0(t)$ with a deterministic trend (*.1 in Example 3.2.2.1) with estimated parameters (*.3 in Example 3.2.2.1):

$$\hat{m}(t) = 0.9999 \sin(2\pi t)$$
(*.9)

Similarly to (*.3) we calculate:

t	0	1/8	2/8	3/8	4/8	5/8	6/8	7/8	1
$\varepsilon(t)$	0.0	1.0	0.0001	−1.0	0.0	1.0	−0.0001	−1.0	0.0

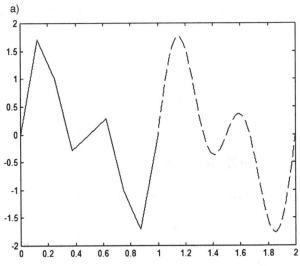

Fig. 3.26a Forecasting with model (*.8) from Example 3.2.2.4

b)

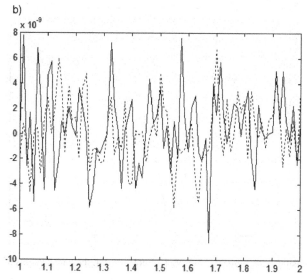

Fig. 3.26b Forecasting with model (*.8) from Example 3.2.2.4: two realizations of the stationary stochastic process $Z_0(t)$

and as in (*.4)–(*.7) we obtain

$$\bar{Z}_0 = \frac{1}{9}\sum_{i=1}^{9} Z_0(t_i) = \frac{1}{9}\sum_{i=1}^{9} \varepsilon(t_i) = 0.0 \qquad (*.10)$$

$$\hat{c}_0 = 0.44,$$
$$\hat{\rho}_0 = 1, \hat{\rho}_1 = -0.0, \hat{\rho}_2 = 0.75 \qquad (*.11)$$

With an analogous assumption as in (*.8) we use the following model for forecasting:

$$Z(t) = 0.9999 \sin(2\pi t) + Z_0(t),$$
$$Z_0(t) \sim N(0, \hat{c}_0) = N(0, 0.44) \qquad (*.12)$$

Results of this forecasting can be seen in Fig. 3.27.

Overview: As one can see, there are enough different approaches to help "discover" a hidden structure in seeming chaotic data. Depending on the final purpose of the data analysis, we can consider deterministic or stochastic approaches. It should be noted that the deterministic way is a "simple" one, based on fewer additional theoretical, model-dependent assumptions, for it is nearly impossibly for real data to satisfy all these assumptions. The stochastic way is more complicated, but an elegant and powerful approach if the fulfillment of all model assumptions is guaranteed. Analyzing a time series is like looking through a keyhole. We can see only a small part of the whole thing. Figure 3.28 demonstrates the limitations of any mathematical model: What happens if the "real structure" of our data out of the given "keyhole" perspective changes completely? We will learn more about useful stochastic approaches in Sect. 3.3.

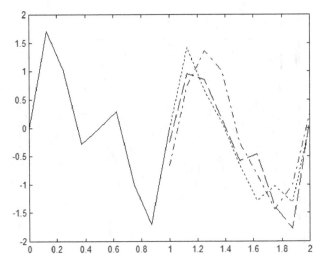

Fig. 3.27 Forecasting with model (*.12) from Example 3.2.2.4: three realizations of the stochastic process

3.3 Considering the Influence of Space, Time, and Other Factors

In this section we discuss some special characteristics that can be used for analyzing spatial or time-induced influences. We have already described a few of them in Sects. 3.1 and 3.2, and for those we only deal with their empirical estimators and give some examples. Other approaches are introduced and explained in a more detailed way, especially by way of some examples.

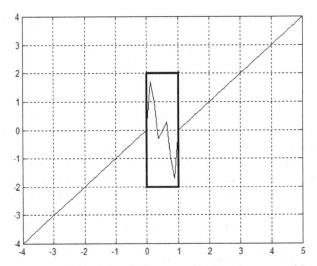

Fig. 3.28 An enlargement of a keyhole: who knows exactly the real structure of data?

3.3.1 The Classical Correlation Coefficient, the Correlation Function, and the Influence Function

The classical *correlation coefficient* $\rho = \rho(X, Y)$ between two random values (or vectors) X and Y is defined as follows:

$$\rho(X, Y) = \frac{Cov(X, Y)}{\sqrt{Var(X)\,Var(Y)}},$$

$$Cov(X, Y) = E(X - EX)(Y - EX),$$

$$Var(X) = Cov(X, X), \quad Var(Y) = Cov(Y, Y)$$

$$(3\text{-}81)$$

The role of the correlation coefficient ρ is similar to that of the cosine between two—here random—vectors X and Y. This coefficient takes values between -1 and 1, and the squared variance corresponds to the length of a random vector. Thus, we can account for the fact that the correlation coefficient helps us to understand and to discover a linear relation between X and Y. If the correlation coefficient is equal to 1, there is a linear relation between X and Y, that is, $Y = c \cdot X, c > 0$. There is also a linear relation between X and Y if the correlation coefficient is equal to -1, but with $Y = c \cdot X, c < 0$. If the correlation coefficient equals zero, we speak about uncorrelated random values X and Y, but not necessarily stochastic independent variables. For $0 < |\rho| < 0.4$ it is usual to speak about weak linearity, for $0.4 \le |\rho| < 0.8$ about middle linearity, and for $|\rho| \ge 0.8$ about strong linearity between X and Y. Sometimes the value 0.7 is used as a lower-limit value instead of 0.8 for strong linearity in statistics texts; it is a matter of choice.

Remark: Mostly mistakes are made interpreting the correlation coefficient as an absolute characteristic describing any kind of possible relations between X and Y. Once again, let us repeat that this coefficient only measures a *linear relation* between these random variables or vectors. Now think about the situation in which a nonlinear relation between X and Y is given following an analytical function $f : Y = f(X)$. If the functional relation f between X and Y is reversible, then a new variable denoted as Y^{-1} exists with $Y^{-1} = f^{-1}(Y)$. If a linear relation between X and the variable Y^{-1} is found by considering the correlation coefficient $\rho(X, Y^{-1})$, then there is a nonlinear relation $f : Y = f(X)$ between X and Y.

The *empirical correlation coefficient* plays an important role in real applications because in most cases the random values X and Y with their theoretical distributions are not given, but finite measurements $x_1, \ldots, x_N$ and $y_1, \ldots, y_N$ of the two variables are available. The definition of the empirical correlation coefficient is as follows:

$$\hat{\rho}(x, y) = \frac{s_{xy}}{\sqrt{s_{xx} \cdot s_{yy}}},$$

$$s_{xy} = \sum_{i=1}^{N} (x_i - \bar{x})(y_i - \bar{y}), \; s_{xx} = \sum_{i=1}^{N} (x_i - \bar{x})^2, \; s_{yy} = \sum_{i=1}^{N} (y_i - \bar{y})^2, \; \bar{x} = \frac{1}{N}\sum_{i=1}^{N} x_i, \; \bar{y} = \frac{1}{N}\sum_{i=1}^{N} y_i$$

$$(3\text{-}82)$$

Obviously, the empirical correlation coefficient tests in an analogous way whether or not a linear relation between the variables being considered is given. We demonstrate this in Example 3.3.1.1

Example 3.3.1.1 Let the following data be given:

x	−0.5	−0.4	−0.3	−0.2	−0.1	0.0	0.1	0.2	0.3	0.4	0.5
y	0.183	0.165	0.102	−0.006	0.058	0.048	0.009	0.053	0.097	0.153	0.279

Simple plotting of x and y as in Fig. 3.29 shows that the relation between the variables is probably nonlinear. Indeed, calculating the empirical coefficient $\rho\,(x,\ y)$ we obtain:

$$\bar{x} = \frac{1}{11}(-0.5 + (-0.4) + \ldots + 0.4 + 0.5) = 0,$$

$$\bar{y} = \frac{1}{11}(0.183 + 0.165 + \ldots + 0.153 + 0.279) = 0.104$$

$$\hat{\rho}\,(x,\ y) = \frac{(-0.5 - 0)\,(0.183 - 0.104) + \ldots + (0.5 - 0)\,(0.279 - 0.104)}{\sqrt{\left[(-0.5 - 0)^2 + \ldots + (0.5 - 0)^2\right] \cdot \left[(0.183 - 0.104)^2 + \ldots + (0.279 - 0.104)^2\right]}} = 0.17$$

$$(*.1)$$

From a more detailed consideration it follows that the relation $y = x^2$ can be assumed, and if this assumption holds, new measurements can be constructed by

$$y_i^{-1} = \begin{cases} -\sqrt{|y_i|}, & x_i < 0 \\ \sqrt{|y_i|}, & x_i \geq 0 \end{cases}, \quad i = 1 \ldots N \qquad (*.2)$$

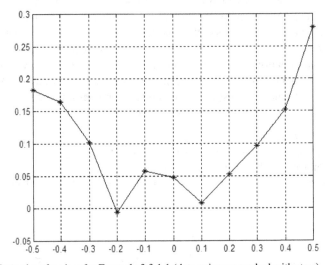

Fig. 3.29 Data plot of y via x for Example 3.3.1.1 (data pairs are marked with stars)

which corresponds in our case to

y^{-1}	−0.43	−0.41	−0.32	−0.077	−0.24	0.22	0.09	0.23	0.31	0.39	0.53

Calculating the new measurements, we find that the inverse function of f consists of two different functions with respect to the given intervals of x, which proves the linearity between x and y^{-1}.

Now we obtain the following results:

$$\bar{x} = \frac{1}{11}(-0.5 + (-0.4) + \dots + 0.4 + 0.5) = 0,$$

$$\bar{y}^{-1} = \frac{1}{11}(-0.43 + (-0.41) + \dots + 0.39 + 0.53) = 0.03$$

$$\hat{\rho}(x, y^{-1}) = \frac{(-0.5 - 0)(-0.43 - 0.03) + \dots + (0.5 - 0)(0.53 - 0.03)}{\sqrt{\left[(-0.5 - 0)^2 + \dots + (0.5 - 0)^2\right] \cdot \left[(-0.43 - 0.03)^2 + \dots + (0.53 - 0.03)^2\right]}} = 0.96$$

$$(*.3)$$

Owing to the strong linearity between X and Y^{-1} indicated by the empirical correlation coefficient equaling 0.96, a functional relation of the kind $Y = X^2$ between the original data should be accepted.

One can test the degree of linearity using the following hypothesis:

$$H_0: \quad \rho(X, Y) = \rho_0 \tag{3-83}$$

The parameter ρ_0 is a given value between -1 and 1. R. A. Fisher proposed using the following test value:

$$T = \frac{w - \bar{w}}{s_W},$$

$$w = \frac{1}{2}\ln\left(\frac{1 + \hat{\rho}(x, y)}{1 - \hat{\rho}(x, y)}\right), \quad \bar{w} = \frac{1}{2}\ln\left(\frac{1 + \rho_0}{1 - \rho_0}\right) + \frac{\rho_0}{2(n-1)}, \quad s_W = \sqrt{\frac{1}{n-3}} \tag{3-84}$$

Here, the estimation of the correlation coefficient from (3-82) is used. It is recommended to take $n > 24$ observations of the parameters X and Y into account. The hypothesis H_0 should be rejected with an error α of type I if

$$|T| \geq z_{1-\alpha/2} \tag{3-84'}$$

where z_q is the q-quantile of a normally distributed random variable.

Remark: The strong linearity between X and Y does not explain the reasons, the roots of this linearity. What if there is a third parameter Z directly influencing both variables, thus causing the secondary, high dependence between X and Y? In order to prove this possibility, so-called *partial correlation coefficients* should be considered.

Let us regard n observations (x, y, z) of the triple (X, Y, Z), which is assumed to follow a three-dimensional normal distribution. The following characteristics are called *empirical partial correlation coefficients*:

$$\hat{\rho}(x, y/z) = \frac{\hat{\rho}(x, y) - \hat{\rho}(x, z) \cdot \rho(y, z)}{\sqrt{(1 - \hat{\rho}^2(x, z))(1 - \hat{\rho}^2(y, z))}},$$

$$\hat{\rho}(x, z/y) = \frac{\hat{\rho}(x, z) - \hat{\rho}(x, y) \cdot \rho(y, z)}{\sqrt{(1 - \hat{\rho}^2(x, y))(1 - \hat{\rho}^2(y, z))}}, \tag{3-85}$$

$$\hat{\rho}(y, z/x) = \frac{\hat{\rho}(y, z) - \hat{\rho}(x, z) \cdot \rho(x, y)}{\sqrt{(1 - \hat{\rho}^2(x, z))(1 - \hat{\rho}^2(x, y))}}$$

The estimators of "usual" correlation coefficients are given in (3-82).

In order to test the hypothesis about the independence of X and Y after removing the primary influence of Z, the following test value should be calculated:

$$T = \sqrt{n-3} \frac{\hat{\rho}(x, y/z)}{\sqrt{1 - \hat{\rho}^2(x, y/z)}} \tag{3-86}$$

The hypothesis about independence is rejected with an error α of type I if

$$|T| \geq t_{m;1-\alpha/2}, \quad m = n - 3 \tag{3-86'}$$

Here, with $t_{m;q}$ the q-quantile of Student's t-distribution with m degrees of freedom is denoted.

Example 3.3.1.2 We consider the following 34 three-dimensional observations and test the independence of X and Y after removing the primary influence of Z:

Num.	1	2	3	4	5	6	7	8	9	10
x	0.45	0.47	0.72	0.07	0.19	0.34	0.42	0.88	0.29	0.64
y	0.51	0.50	0.74	0.3	0.22	0.31	0.56	0.76	0.45	0.55
z	0.48	0.47	0.16	0.54	0.77	0.67	0.3	0.0	0.38	0.2

11	12	13	14	15	16	17	18	19	20
0.0	0.63	0.79	0.0	0.60	0.51	0.61	0.48	0.5	0.51
0.0	0.55	0.86	0.2	0.63	0.52	0.62	0.51	0.74	0.56
1.0	0.26	0.08	0.74	0.14	0.26	0.38	0.27	0.21	0.26

21	22	23	24	25	26	27	28	29	30
0.41	0.58	0.27	0.17	0.41	0.14	0.20	0.45	0.50	0.11
0.52	0.58	0.27	0.35	0.40	0.24	0.48	0.56	0.45	0.39
0.27	0.25	0.57	0.51	0.43	0.63	0.33	0.23	0.22	0.44

31	32	33	34
0.11	0.26	0.37	0.19
0.41	0.43	0.49	0.41
0.42	0.37	0.30	0.43

For a primary analysis of the data, we recommend visualizing these values as shown in Fig. 3.30. First, the "usual" empirical correlation coefficients are calculated, for which we use equation (3-82):

$$\hat{\rho}(x, y) = 0.87,$$
$$\hat{\rho}(x, z) = 0.81, \tag{*.1}$$
$$\hat{\rho}(y, z) = -0.92$$

As one can see, there is strong linearity among all the parameters. Using (3-85), we find that the empirical partial correlation coefficient between X and Y after removing influence of Z corresponds to

$$\hat{\rho}(x, y/z) = \frac{\hat{\rho}(x, y) - \hat{\rho}(x, z) \cdot \rho(y, z)}{\sqrt{(1 - \hat{\rho}^2(x, z))(1 - \hat{\rho}^2(y, z))}}$$
$$= \frac{0.87 - 0.81 \cdot (-0.92)}{\sqrt{(1 - 0.81^2)(1 - 0.92^2)}} = 0.54 \tag{*.2}$$

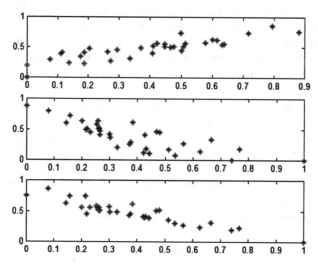

Fig. 3.30 Visualization of the data for primary analysis: data clouds (x, y), (x, z), and (y, z)

The empirical partial correlation coefficient indicates middle linearity between X and Y after removing the influence of Z. But these parameters are not independent because, using (3-86) and (3-86'), the independence hypothesis should be rejected with error $\alpha = 0.05$ of type I:

$$T = \sqrt{n-3} \, \frac{\hat{\rho}\,(x,\,y/z)}{\sqrt{1-\hat{\rho}^2\,(x,\,y/z)}}$$

$$= \sqrt{34-3} \, \frac{0.54}{\sqrt{1-0.54^2}} = 3.59 \,, \qquad (*.3)$$

$$t_{34-3;1-0.05/2} = 2.04$$

It can be proved that

$$|T| \geq t_{34-3;1-0.05/2} \qquad (*.4)$$

Thus, removing the influence of Z leads to a significant decrease in the degree of linearity between X and Y, but not to confirmation of the hypothesis about independence between X and Y.

We defined the correlation function and showed its estimation earlier in Sect. 3.2.2. Not every function that looks "optically" like a correlation function actually is one. A correlation function is characterized by certain properties that have to be fulfilled, and it is often difficult to prove all of these properties at the same time. There are a great many models of correlation functions that can be considered for different applications, but fitting these functions is sometimes complicated: for example, from a statistical point of view if the number of datum is too small. Moreover, what if we do not want to assume a stochastic relation between the parameters, that is, correlation? Is there another—"deterministic"—possibility to quantify the influence? Of course there is such a possibility. We now discuss a deterministic equivalent of the correlation function, known as an *influence function*, which an alternative method for analyzing interaction. The method is very simple. The most important advantage of using an influence function instead of a correlation function is that there are no restrictive assumptions.

We explain the basic idea of the influence function method using an example from forestry. We assume that each tree characteristic (e.g., parameter and mark)—DBH (diameter at breast height), tree height, BAI (basal area increment), and others—depends on other trees, namely on their locations and characteristics. It is realistic to propose that the influence of one tree on another decreases with increasing distance between them. We denote the final tree distance where this influence f still exists with the parameter R. Thus, it is true that $f(r) = 0, r > R$. It is assumed that the function f is the same for every tree in a forest or a stand.

The second model assumption concerns the linearity of the average of single influences belonging to neighboring trees, which means for characteristic m of a tree located at point x_0 with N neighboring trees at distances less than R:

$$m(x_0) = m(x_1) \cdot w(r_{10}) + m(x_2) \cdot w(r_{20}) + \ldots + m(x_N) \cdot w(r_{N0}),$$

$$r_{i0} = |x_i - x_0|, i = 1 \ldots N,$$

$$w(r_{i0}) = \frac{f(r_{i0})}{\sum\limits_{i=1}^{N} f(r_{i0})}, \sum\limits_{i=1}^{N} w(r_{i0}) = 1 \qquad (3\text{-}87)$$

The sum of weights is assumed to equal one in order to avoid systematic under- or overestimation However, our aim is to choose this influence function f "optimally." A polynomial function

$$f(r) = 1 + a_1 t + \ldots + a_M t^M \qquad (3\text{-}88)$$

can be taken into account. We could also use another functional model. The method of least squares can be applied for parameter fitting. Thus, the following sum has to be minimized:

$$\sum_{\forall(x_0, N)} \left[\frac{m(x_1) \cdot f(r_{10}) + m(x_2) \cdot f(r_{20}) + \ldots + m(x_N) \cdot f(r_{N0})}{\sum\limits_{i=1}^{N} f(r_{i0})} - m(x_0) \right]^2 \rightarrow \min$$

$$(3\text{-}89)$$

The sum in (3-89) considers all the trees in the forest. It should be noted that the number of neighbors N can vary for different trees. Further, equation (3-89) can be transformed to

$$\sum_{\forall(x_0, N)} \frac{\left[m(x_1) \cdot f(r_{10}) + m(x_2) \cdot f(r_{20}) + \ldots + m(x_N) \cdot f(r_{N0}) - m(x_0) \sum\limits_{i=1}^{N} f(r_{i0}) \right]^2}{\left[\sum\limits_{i=1}^{N} f(r_{i0}) \right]^2} \rightarrow \min$$

$$(3\text{-}89')$$

An alternative presentation of (3-87) is

$$m(x_0) \cdot \sum_{i=1}^{N} f(r_{i0}) = m(x_1) \cdot f(r_{10}) + m(x_2) \cdot f(r_{20}) + \ldots + m(x_N) \cdot f(r_{N0}),$$

$$r_{i0} = |x_i - x_0|, i = 1 \ldots N$$

$$(3\text{-}87')$$

Thus, it is sufficient to minimize

$$\sum_{\forall(x_0, N)} \left[m(x_1) \cdot f(r_{10}) + m(x_2) \cdot f(r_{20}) + \ldots + m(x_N) \cdot f(r_{N0}) - m(x_0) \sum_{i=1}^{N} f(r_{i0}) \right]^2 \rightarrow \min$$

$$(3\text{-}90)$$

Using (3-88) in (3-90) and applying least-squares techniques, we get a LSE (linear system of equations) with respect to the unknown parameters $a_1, \ldots, a_M$:

$$a_1 \sum_{\forall(x_0,\,N)} \left[\sum_{i=1}^{N} m(x_i) r_{i0} - m(x_0) \sum_{i=1}^{N} r_{i0} \right] \left[\sum_{i=1}^{N} m(x_i) r_{i0}^j - m(x_0) \sum_{i=1}^{N} r_{i0}^j \right] +$$

$$a_2 \sum_{\forall(x_0,\,N)} \left[\sum_{i=1}^{N} m(x_i) r_{i0}^2 - m(x_0) \sum_{i=1}^{N} r_{i0}^2 \right] \left[\sum_{i=1}^{N} m(x_i) r_{i0}^j - m(x_0) \sum_{i=1}^{N} r_{i0}^j \right] +$$

$$\cdots$$

$$a_M \sum_{\forall(x_0,\,N)} \left[\sum_{i=1}^{N} m(x_i) r_{i0}^M - m(x_0) \sum_{i=1}^{N} r_{i0}^M \right] \left[\sum_{i=1}^{N} m(x_i) r_{i0}^j - m(x_0) \sum_{i=1}^{N} r_{i0}^j \right] =$$

$$- \sum_{\forall(x_0,\,N)} \left[\sum_{i=1}^{N} m(x_i) - m(x_0) \cdot N \right] \left[\sum_{i=1}^{N} m(x_i) r_{i0}^j - m(x_0) \sum_{i=1}^{N} r_{i0}^j \right], \quad j = 1 \dots M$$

$$(3\text{-}91)$$

It is meaningful to choose the degree of polynom M to be far smaller than the number of trees in the forest. This measure, the sum of differences from (3-90), can be used in the second step for fitting the best value of the parameters R and M discussed above. If there are n trees, the accuracy σ of fitting the influence function can be obtained by

$$\sigma^2 = \frac{1}{n} \sum_{\forall(x_0,\,N)} \left[m(x_1) \cdot f(r_{10}) + m(x_2) \cdot f(r_{20}) + \dots + m(x_N) \cdot f(r_{N0}) - m(x_0) \sum_{i=1}^{N} f(r_{i0}) \right]^2,$$

$$(3\text{-}90')$$

An example for fitting the influence function follows.

Example 3.3.1.3 We fit an influence function for the following data set (see Fig. 3.31):

	1	2	3	4	5	6	7	8	9
x	0.67	0.81	3.90	5.02	5.10	3.87	5.66	6.35	6.98
y	1.62	2.48	3.78	4.02	3.35	2.05	3.46	4.65	3.74
m	3.23	3.31	2.00	1.06	2.90	1.30	0.93	2.74	2.50

10	11	12	13	14	15	16	17	18	19
8.17	6.74	6.72	8.12	8.66	1.42	0.89	1.40	3.93	3.24
4.37	3.04	1.77	0.65	2.66	5.46	7.48	9.55	9.06	7.77
1.94	1.17	1.66	3.27	2.05	2.62	2.50	2.30	1.95	1.40

20	21	22	23	24	25	26	27	28	29
2.59	2.92	3.72	4.59	5.46	6.77	7.46	8.92	8.54	7.80
6.58	5.10	5.47	7.07	6.58	6.56	8.50	9.44	8.54	6.16
3.16	1.74	1.83	1.33	2.00	1.45	3.75	2.91	2.46	2.34

Here (x, y) denotes the tree location and m is a tree parameter: BHD, BAI, or another characteristic. We first calculate the distance matrix d between the trees in order to determine the number of neighbors of each tree. For example, for the first tree—the first column in this matrix with column numeration denoted with ":"—we obtain the following tree-tree distances:

	1	2	3	4	5	6	7	8	9
$d(1, :)$	0.0	0.87	3.89	4.96	4.75	3.23	5.32	6.44	6.66

10	11	12	13	14	15	16	17	18	19
7.99	6.23	6.05	7.51	8.06	3.91	5.87	7.96	8.12	6.66

20	21	22	23	24	25	26	27	28	29
5.32	4.14	4.91	6.71	6.90	7.85	9.67	11.37	10.49	8.45

Let us set the parameter R equal to 5, so the first tree has eight neighboring trees. The distance to the trees with numbers 2–6, 5, 21, 22 is smaller than R. We first assume that the influence function is a polynomial of degree five, which means

$$f(r) = 1 + a_1 t + \ldots + a_5 t^5 \qquad (*.1)$$

After solving (3-91) we get

$$(a_1, \ldots, a_5) = (-2.5, 2.13, -0.8, 0.14, -0.01) \qquad (*.1')$$

Figure 3.32 shows this polynomial. The fitting accuracy σ from (3-90′) corresponds to

$$\sigma = 0.066 \qquad (*.2)$$

The minimum of the influence function marks the tree distance that is harmful for increasing of the tree parameter m. It corresponds here to the distance 1.2.

Second, we test a polynom of degree eight for influence function fitting and get

$$f(r) = 1 + a_1 t + \ldots + a_8 t^8 \qquad (*.3)$$

Solving (3-91) yields

$$(a_1, \ldots, a_8) = (-4.19, 6.93, -6.12, 3.19, -1.01, 0.19, -0.02, 0.001) \qquad (*.3')$$

Figure 3.33 shows this polynom. The fitting accuracy σ from (3-90′) corresponds to

$$\sigma = 0.01 \qquad (*.4)$$

The minimum of influence function marks the tree distance that is harmful for increasing the tree parameter m, which corresponds here to the distance 0.9.

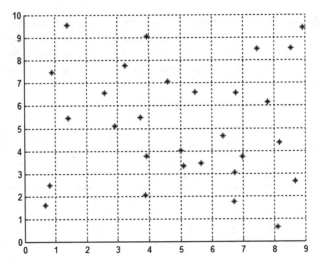

Fig. 3.31 Plot of tree locations (x, y)

A comparison of the accuracies in (*.2) and (*.4) indicates that the polynomial of degree eight fits the influence function better than that of degree five. It is clear that we can optimize this degree value numerically using loops for the procedure (3-91) controlling the accuracy in (3-90′) in a stepwise fashion.

Remark: We have presented some methods that help us to consider different kinds of influence: (1) the relation of one parameter to another (correlation coefficient),

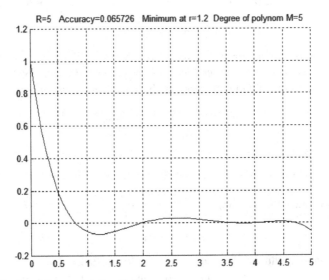

Fig. 3.32 Fit of the influence function (*.1)

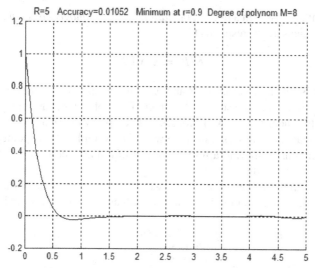

Fig. 3.33 The influence function fitting (*.3)

(2) the dependence of two parameters on a third (partial correlation coefficient), and (3) the dependence of a parameter on time or space (correlation function, influence function). We have noted that there is not only a linear dependence of one parameter on another and have shown a way to help us to determine another kind of dependence using the correlation coefficient (see Example 3.3.1.1). Generally, we have presented two ways or methods—the stochastic and the deterministic approach—and have discussed the arguments for their use.

3.3.2 Points and Marks: Pair and Mark Correlation Function and Point Process Statistics

Point process theory, theory of marked point processes, and stochastic geometry are relatively new mathematical frameworks, which deal with stochastic models for spatially distributed parameters or variables. In particular, such models propose some effective approaches that can be applied in material sciences, geology, biology, forestry, and other fields with discrete spatial structures.

A point process is a mathematical model for describing a random point pattern. A detailed introduction to the theory of point processes can be found in Kallenberg (1986), Karr (1986), and Stoyan and Stoyan (1994). In what follows we provide some useful definitions and explain the corresponding applications.

The intensity measure describes the mean number of random points in a set B in Euclidean space. For the applications discussed below we restrict ourselves to random point processes in a plane. For *stationary and homogeneous point processes* the intensity measure corresponds to

$$\Lambda(B) = \lambda \cdot v(B) \tag{3-92}$$

where $v(B)$ is the area of the set B and λ is the mean number of points in a unit square called intensity. The so-called characteristics of second order play an important role in the theory of marked processes. Each point in a marked point process gets a mark that characterizes it in some way. For example, let the point process be given by locations of trees in a forest. Some useful marks describing a tree are its height, its age, and its diameter at breast height.

We start with the following measure, which describes the mean of a function f taking its values at pairs of random points distributed in two sets B_1 and B_2:

$$\alpha_f^{(2)}(B_1 \times B_2) = E \sum_{[x_1, m_1] \in \Psi} \sum_{\substack{[x_2, m_2] \in \Psi \\ x_2 \neq x_1}} f(m_1, m_2) 1_{B_1}(x_1) 1_{B_2}(x_2) ,$$

$$1_B(t) = \begin{cases} 1, & t \in B \\ 0, & otherwise \end{cases} \tag{3-93}$$

$$t = x_1, x_2$$

If a density function $\rho_f^{(2)}(x_1, x_2)$ of measure (3-93) exists, it is called *f-product density*, and

$$\alpha_f^{(2)}(B_1 \times B_2) = \iint_{B_1 B_2} \rho_f^{(2)}(x_1, x_2) dx_1 dx_2 \tag{3-94}$$

A more "user-friendly" description of the mean of the f-product density $\rho_f^{(2)}(x_1, x_2)$ is the following: We consider two infinitely small discs with centers at random points x_1 and x_2 and with areas dF_1 and dF_2. We define a random variable that takes the value $f(m_1, m_2)$ if exactly one random point is placed in each of these discs. Otherwise this random variable gets the value zero. Then the mean of this random variable is $\rho_f^{(2)}(x_1, x_2) dF_1 dF_2$.

A point process is called isotropic if the following is true:

$$\rho_f^{(2)}(x_1, x_2) = \rho_f^{(2)}(r), \quad r = \|x_2 - x_1\| \tag{3-95}$$

Equation (3-95) means that the f–product density depends on the distance between points and not on their spatial locations. The last but not the least important characteristic of second order is the so-called *mark correlation function* $k_f(r)$, defined as

$$k_f(r) = \frac{\rho_f^{(2)}(r)}{\rho_1^{(2)}(r)} \cdot \frac{\rho_1^{(2)}(\infty)}{\rho_f^{(2)}(\infty)} = c \cdot \frac{\rho_f^{(2)}(r)}{\rho_1^{(2)}(r)} ,$$

$$c = \frac{\rho_1^{(2)}(\infty)}{\rho_f^{(2)}(\infty)}, \quad \rho_1^{(2)}(r) \neq 0, \tag{3-96}$$

$$\rho_1^{(2)}(r) = \rho_f^{(2)}(r) \quad mit \quad f(m_1, m_2) = 1$$

The function $\rho_1^{(2)}(r)$ in (3-96) is called the *pair correlation function*. There are different estimators of the f-product density, and they generally have the following form:

$$\hat{\rho}_f(r) = \sum_{\substack{i,\,j=1,\ldots n \\ i<j}} f(m_i,\,m_j) \cdot w_{ij}(r)\,,$$

$$w_{ij} = \frac{G_{ij}(r)}{\sum_{i,\,j=1,\ldots n} G_{ij}(r)},\qquad G_{ij}(r) \propto \left|r - \|x_j - x_i\|\right| \tag{3-97}$$

where the sign $\propto$ stands for "proportional to." The number of points in a realization of a point process is denoted by n. Stoyan and Stoyan (1992) propose using the following weights $G_{ij}(r)$:

$$G_{ij}(r) = \frac{e_h\left(r - \|x_i - x_j\|\right)}{v\left(W_i \cap W_j\right)},$$

$$e_h(t) = \begin{cases} \dfrac{3}{4h}\left(1 - \dfrac{t^2}{h^2}\right), & -h \le t \le h \\ 0, & otherwise \end{cases} \tag{3-98}$$

In (3-98) the so-called Epanechnikov function $e_h(t)$ is used. The choice of a window W and parameter h is discussed in Stoyan and Stoyan (1994). Usually, one tests different parameters until meaningful results are obtained. Alternative weights are

$$G_{ij}(r) = 1\left(r - h \le \|x_j - x_i\| \le r + h\right) \tag{3-99}$$

This representation of weights is mostly applied for calculating usual histograms, and the estimator of the f-product density with weights from (3-99) can be considered as a special generalized histogram.

Usual forms of function f are

$$\begin{aligned} a) \quad & f(m_i,\,m_j) = m_i \cdot m_j, \\ b) \quad & f(m_i,\,m_j) = |m_i - m_j| \end{aligned} \tag{3-100}$$

where $m_i,\,i = 1 \ldots n$ represents the parameter or mark corresponding to the point $x_i,\,i = 1 \ldots n$. An estimation of the mark correlation function can help us analyze interaction among points within the point process. Once again, think about a point process given by tree positions. If neighboring trees support one another the mark tree height can take relatively large values at each point. In a case of suppression one high tree interferes with the growth of surrounding trees, leading to small values of the mark. If form (b) from (3-100) is used, then we can consider the following estimator of the mark correlation function:

$$\hat{k}(r) = \frac{\hat{\rho}_f(r)}{\hat{m}^{(2)}},$$

$$\hat{m}^{(2)} = \left(\frac{1}{n}\sum_{k=1}^{n} m_k\right)^2 \tag{3-100'}$$

with $\hat{\rho}_f(r)$ from (3-97).

Remark: The forms (a) and (b) from (3-100) stand for different situations that have to be analyzed. For example, (a) leads to the "general influence" of the marks depending on distances between corresponding points. The distance r with a local minimum of $\hat{\rho}_f(r)$ represents a so-called "negative synergy" between the marks. Or, in other words, the mean product of both marks for this distance decreases. In contrast to this fact, the distance r with a local maximum of $\hat{\rho}_f(r)$ represents "positive synergy" between the parameters. Form (b) leads to local minima of $\hat{\rho}_f(r)$ at locations with approximately identical or closely similar marks. Local maxima correspond to locations with increasing differences between marks.

Such interpretations should be handled with care because there are also numerical effects generated by the estimation process. An advantage is the fact that the estimator (3-97) can be applied with or without any assumption about the randomness of the point locations. In deterministic cases this estimator can be interpreted as one for a parameter influence function.

In Example 3.3.2.1 we show in a more detailed way how an estimation of the mark correlation function can be carried out.

Example 3.3.2.1 (continued below) We consider the following marked point pattern with $n = 76$. Once again, think about tree locations at points (x,y) with the corresponding mark m standing for the tree diameter at breast height (DBH) in centimeters:

Number		1	2	3	4	5	6	7	8	9	10
	x	2.68	3.22	15.62	20.06	20.39	15.49	22.64	25.41	27.93	32.69
	y	6.48	9.92	15.14	16.07	13.39	8.20	13.84	18.60	14.98	17.47
	m	64.5	66.2	40.0	21.2	58.0	25.9	18.5	54.8	50.0	38.7

11	12	13	14	15	16	17	18	19	20
26.94	26.87	32.46	34.65	40.95	40.76	48.77	56.63	50.41	45.62
12.18	7.08	2.58	10.64	12.13	16.06	17.91	10.56	7.31	3.20
23.3	33.1	65.4	41.0	53.7	55.1	48.20	59.9	43.8	53.0

21	22	23	24	25	26	27	28	29	30
58.11	63.63	63.98	70.58	72.51	76.3	79.17	78.31	72.5	86.7
1.14	3.41	11.05	10.58	19.04	14.37	15.36	8.62	3.21	9.85
48.3	56.2	47.8	51.5	35.8	26.0	53.1	47.5	60.8	42.2

31	32	33	34	35	36	37	38	39	40
85.6	97.02	98.55	90.82	5.67	3.57	5.6	15.72	12.96	10.33
15.55	12.78	2.94	3.52	21.84	29.93	38.19	36.23	31.07	26.33
24.5	57.7	55.5	55.4	52.3	49.9	46.0	38.9	27.9	63.2

41	42	43	44	45	46	47	48	49	50
11.67	14.86	18.35	21.84	27.08	29.85	35.67	34.18	31.21	40.0
20.37	21.88	28.29	26.34	26.25	34.01	37.77	34.18	24.64	30.33
34.7	36.5	26.6	39.9	29.0	75.0	58.2	49.9	46.8	85.6

51	52	53	54	55	56	57	58	59	60
46.88	46.44	46.51	51.33	59.76	58.71	58.38	63.0	61.07	67.61
38.21	26.86	23.19	29.99	22.5	34.05	38.97	38.88	29.73	32.42
73.1	20.9	50.7	41.2	44.0	38.9	22.5	43.0	38.2	52.6

61	62	63	64	65	66	67	68	69	70
72.35	74.44	74.05	79.96	83.66	84.97	91.93	93.63	89.37	88.03
37.0	33.99	26.08	27.54	27.78	24.53	20.37	28.65	32.9	38.55
43.5	20.4	41.1	47.9	40.5	44.0	46.6	48.4	48.6	61.4

71	72	73	74	75	76
94.29	98.41	99.19	98.77	97.66	78.46
39.67	32.94	25.58	23.58	21.11	36.18
43.1	39.4	36.8	20.7	39.2	50.6

Figure 3.34 presents the data for an observation window. The value $\hat{m}^{(2)}$ from (3-100′) is calculated by

$$\hat{m}^{(2)} = \left(\frac{1}{76} [64.5 + 66.2 + \ldots + 39.2 + 50.6] \right)^2 = 2042 \qquad (*.1)$$

Before the mark correlation function $k_f(r)$ with $f(m_i, m_j) = m_i \cdot m_j$, i, $j = 1 \ldots 76$ is estimated, the distance matrix

$$(d_{ij})_{i, j=1\ldots76}, d_{ij} = \sqrt{(x_i - x_j)^2 + (y_i - y_j)^2}$$

should be determined. After dividing the r-axes into equidistant intervals of width 2Δ, the point pairs $(x_i, y_i), (x_j, y_j), i, j = 1 \ldots 76$ are assigned to an interval $[r_0 - \Delta, r_0 + \Delta]$ by testing $r_0 - \Delta \le d_{ij} < r_0 + \Delta$. If the weighting method with kernel function (3-99) is used, we should summarize the products of marks for point pairs belonging to each interval and divide the final sum by the number of such pairs. With (3-97) and (3-100′) the estimation $\hat{k}_f(r)$ is completed. Figure 3.35 shows the empirical mark correlation function for the given data. It can be seen, that there is a

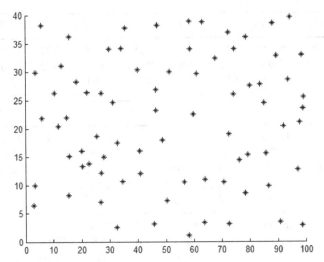

Fig. 3.34 Point pattern of Example 3.3.2.1 in an observation window 100×40 square meters

local minimum at about $r = 4$ meters. Probably, there is "negative synergy" among neighboring trees at small distances. This fact could lead to smaller DBH marks of such trees. However, there is no significant correlation for tree distances greater than 8 meters. The statistical analysis of these data is continued below.

Sometimes it is interesting to know the type of point process to which the given point pattern belongs. We start by discussing so-called *Poisson point fields*.

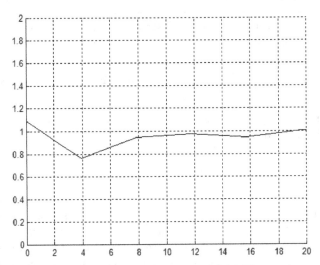

Fig. 3.35 Empirical mark correlation function for Example 3.3.2.1. Distances between points r are given in meters

Poisson point fields or more briefly Poisson fields are the simplest and most often applied forms of point fields because some useful characteristics for them can be calculate exactly. The term "Poisson" indicates the distribution of point numbers. In particular, the point numbers in disjoint (nonoverlapping) sets are stochastically independent. Poisson fields are used for constructing more complicated models such as *Neyman-Scott cluster fields*. In this case, the points are divided into two classes, known as parent and daughter points. The first come from the initial Poisson process, and in the second step the daughter points are scattered around each parent point like seeds around an "old" tree.

3.3.2.1 Homogeneous Poisson fields

One speaks about a field being a *homogeneous Poisson field* if it has the following properties:

1. The random number of points $N(S_1), \ldots, N(S_n)$ in disjoint sets $S_1, \ldots, S_n$ are stochastically independent, which means that

$$P(N(S_1) = k_1, \ldots, N(S_n) = k_n) = P(N(S_1) = k_1) \cdot \ldots \cdot P(N(S_n) = k_n)$$

2. The random number of points $N(S)$ in a set follows a Poisson distribution with parameter $\lambda v(S)$. With $v(S)$ the volume (area) of the set is given. Thus, λ indicates the mean of the random number of points in a set with volume equal to one. The Poisson distribution of the number of points corresponds to

$$P(N(S) = k) = \frac{[\lambda v(S)]^k}{k!} \exp(-\lambda v(S)).$$

For the simulation of homogeneous Poisson fields the following property is important: If a set includes exactly n points, then these points are uniformly and independently distributed in this set.

The independence properties (1) and (2) described above make it possible to calculate certain conditional probabilities. One of these leads to an important function called the D-function or, more precisely, $D(r)$, which describes the distribution of the mean nearest-neighbor distance r for a point of the point process. We discuss a statistical estimator for this function in order to better explain its meaning. If n points located at $x_1, \ldots, x_n$ are given in an observation window, an estimator $\hat{D}(r)$ of the function $D(r)$ can be calculated as follows:

$$\hat{D}(r) = \frac{1}{n} \sum_{i=1}^{n} 1\,(0 \leq D_i \leq r),$$

$$D_i = \min(|x_i - x_1|, \ldots, |x_i - x_{i-1}|, |x_i - x_{i+1}|, \ldots, |x_i - x_n|), \quad i = 1, \ldots, n$$

$$(3\text{-}101)$$

In (3-101) the indication D_i, $i = 1, \ldots, n$ describes the distance to the nearest neighbor of the point located at x_i, $i = 1, \ldots, n$. In the sum the cases are summarized, where these distances are less than or equal to r.

The theoretical calculation of the function $D(r)$ for homogeneous Poisson point fields leads to

$$D(r) = 1 - \exp\left(-\lambda \pi r^2\right), r \geq 0 \qquad (3\text{-}102)$$

The density of this function corresponds to

$$d(r) = 2\lambda \pi \exp\left(-\lambda \pi r^2\right), r \geq 0 \qquad (3\text{-}103)$$

Further details about determining (3-102) and (3-103) can be found in Stoyan and Stoyan (1994).

Another important characteristic is the so-called *K-function*. Its exact definition and mean come from $\lambda K(r)$, indicating the mean number of points in a disc of radius r centered at a point of the point field. This point in the center of this disc is not counted. For homogeneous Poisson field the following holds [see Stoyan and Stoyan (1994)]:

$$K(r) = \pi r^2, r \geq 0 \qquad (3\text{-}104)$$

For some applications it is meaningful to standardize the function K leading to the *L-function*, which is defined as

$$L(r) = \sqrt{\frac{K(r)}{\pi}}, r \geq 0 \qquad (3\text{-}105)$$

For homogeneous Poisson fields this leads to (see Fig. 3.36):

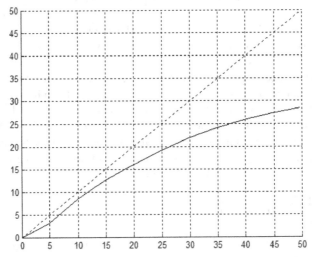

Fig. 3.36 Empirical *L*-function for the data from Example 3.3.2.1 (*solid*) and *L*-function of a Poisson field (*dashed*)

$$L(r) = r, \ r \geq 0 \tag{3-105'}$$

The simulation of Poisson fields is an important task because it is applied in field statistics and provides a "starting point" for simulating more complicated structures: for example, Neyman-Scott cluster processes. Currently, there are many software tools that support point field simulation. Details about the algorithm of this simulation can be found in Stoyan and Stoyan (1994).

Now we discuss some statistical approaches developed for homogeneous Poisson fields. "Statistical" here means that we have a realization—a set of scattered points—of an unknown point process and want to estimate some of its characteristics. It should be noted that here, unlike in "classical statistics" but similar to geostatistics, a single realization is taken for estimation. Thus, some constraints should be imposed in order to get an estimator. If we assume a homogeneous field, we can estimate its intensity in an observation window W by the quotient of point numbers in the window $N(W)$ and the area of the window $v(W)$:

$$\hat{\lambda} = \frac{N(W)}{v(W)} \tag{3-106}$$

For inhomogeneous fields equation (3-106) is no longer an allowed method, but there are other ways to estimate the intensity function [see Stoyan and Stoyan (1994)].

There are various methods that allow us to determine whether or not the point muster belongs to a realization of a homogeneous Poisson field. We discuss two groups of methods here. The first describes so-called *square count methods* and the second is based on the *L*-function. We now show how these methods work.

Group 1: Dispersion Index Method

At first, the observation window is divided in k subregions of equal area. The number of points N_i, $i = 1, \ldots, k$ in each region should be counted. Assuming a Poisson field (hypothesis), these N_i, $i = 1, \ldots, k$ are independently and identically distributed with the mean per region equal to $\lambda v(W)/k$. In order to test this hypothesis, the following characteristic (test value) is calculated:

$$T = \frac{(k-1)s^2}{\bar{N}}, \quad \bar{N} = \frac{1}{k}\sum_{i=1}^{k} N_i, \quad s^2 = \frac{1}{k-1}\sum_{i=1}^{k}(N_i - \bar{N})^2 \tag{3-107}$$

If

$$T > \chi^2_{k-1, \ \alpha} \ or \ \ T < \chi^2_{k-1, \ 1-\alpha} \tag{3-108}$$

then the Poisson hypothesis is rejected with probability α of error of type I.

With $\chi^2_{b, \ a}$ the a-quantile of the χ^2-distribution with b degrees of freedom is denoted. Stoyan and Stoyan (1994) recommend using $k > 6$ and $\lambda v(W)/k > 1$. For the rejection case the following two causes can be proposed: Points appear in clusters or their pattern shows more regularity than could be expected for a realization of a homogeneous Poisson field.

Remark: The so-called Greig-Smith test is a refined version of the dispersion index test. More details can be found in Stoyan and Stoyan (1994).

Group 2: *L*-Function Tests

Here the empirical *L*-function should be calculated. Such tests assess the deviation from the distributional properties of a Poisson field in a more sensible way. The test is based on property (3-105). Therefore, as the test value we consider the following:

$$T = \max_{r \le \max(r)} \left| \hat{L}(r) - r \right| \tag{3-109}$$

If the point pattern has n points in the observation window and the assumption about the Poisson character holds, then the following term can be used as an estimator of the *L*-function or as an empirical *L*-function in (3-109):

$$\hat{L}(r) = \frac{1}{\sqrt{\pi}} \sqrt{\frac{1}{n} \sum_{i=1}^{n} \sum_{\substack{j=1 \\ j \ne i}}^{n} 1\,(0 \le r_{ij} \le r)}, \tag{3-109'}$$

$$r_{ij} = \left| x_i - x_j \right|, \quad i, j = 1, \ldots, n$$

If a Poisson character cannot be assumed, the generalized relation between the *L*- and the *K*-function (3-105) should be used. There are further—also unbiased— estimators for $\lambda K(r)$, $\lambda^2 K(r)$ [see Stoyan and Stoyan (1994) for discussions about their advantages and disadvantages]. Practically, the estimation of the *L*-function can go through two steps. First, the intensity and $\lambda K(r)$ are estimated by (3-106) and

$$\lambda \hat{K}(r) = \frac{1}{n} \sum_{i=1}^{n} \sum_{\substack{j=1 \\ j \ne i}}^{n} 1\,(0 \le r_{ij} \le r), \tag{3.109''}$$

$$r_{ij} = \left| x_i - x_j \right|, \quad i, j = 1, \ldots, n$$

Second, the estimator for the *L*-function is determined by (see Fig. 3.36):

$$\hat{L}(r) = \sqrt{\frac{\lambda \hat{K}(r)}{\hat{\lambda} \pi}}, \quad r \ge 0 \tag{3-110}$$

The Poisson hypothesis is rejected with the argument that it is not a realization of a Poisson process if this test value T becomes "too large" or is greater than a "limit value." This limit value can vary for other estimators of L.

If this version of the *L*-test is not acceptable, a *Monte Carlo L-test* can be used: One simulates 999 realizations of a homogeneous Poisson field in an observation window with exactly *n* points as given in the point pattern that has to be tested. For each sample the empirical *L*-functions should be estimated, and the corresponding test values are calculated following (3-109):

$$T_i = \max_{r \leq \max(r)} \left| \hat{L}_i(r) - r \right|, i = 1, \ldots, 999 \tag{3-111}$$

We order these 999 test values and the value *T* calculated for the given point pattern leading to 1000 ordered values:

$$T^{(1)} \leq T^{(2)} \leq \ldots \leq T^{(1000)} \tag{3-111'}$$

If the index of the value *T* belonging to the real pattern in this series (3-111') is greater than 950 or 990 then the Poisson field hypothesis should be rejected with probability α of error of type I.

Remark: Using Monte Carlo tests is popular for goodness-of-fit tasks for any point process.

Example 3.3.2.1 (continued) First, we test the Poisson field hypothesis for the data from Example 3.3.2.1 by the dispersion index method. The observation window in Fig. 3.34 is divided into ten subregions of area 20×20 square meters. The number of points N_i, $i = 1, \ldots, 10$ in each region is counted:

9	6	8	8	11
4	10	7	8	5

In order to test this hypothesis, the following characteristic (test value) is calculated:

$$\bar{N} = \frac{1}{10} \sum_{i=1}^{10} N_i = \frac{1}{10} [9 + 6 + \ldots + 5] = \frac{76}{10} = 7.6, \; s^2 = \frac{1}{10-1} \sum_{i=1}^{10} (N_i - \bar{N})^2 = 4.71$$

$$T = \frac{(10-1)s^2}{\bar{N}} = 5.58.$$

$$(*.2)$$

The Poisson hypothesis is not rejected with probability $\alpha = 0.05$ of error of type I because

$$\chi^2_{10-1, \, 0.05} = 3.3251, \quad \chi^2_{10-1, \, 1-0.05} = 16.919 \quad \Rightarrow$$

$$\chi^2_{10-1, \, 0.05} \leq T \leq \chi^2_{10-1, \, 1-0.05}$$

$$(*.3)$$

We can conclude that the points are not scattered more regularly as expected for a realization of a homogeneous Poisson field.

Second, the empirical L-function is calculated. The intensity can be estimated by (3-106):

$$\hat{\lambda} = \frac{76}{100 \cdot 40} = 0.0204 \qquad (*.4)$$

We use the distance matrix for calculating $\lambda K(r)$ following (3.109″). For this, we calculate the number of neighbors at distances less than r meters for each point. These numbers are cumulated over all data points and divided by the number $n = 76$ of these points. Using (3-110) we get the empirical L-function shown in Fig. 3.36.

The behavior of this function for large r is not really typical for the realization of a Poisson field. Applying the Monte-Carlo L-test based on the calculation of (3-111) and (3-111′) leads to a rejection of the Poisson field hypothesis with probability $\alpha = 0.05$ of error of type I, which confirms the fact that L-tests are more sensitive than dispersion index tests.

Now we discuss some details concerning inhomogeneous Poisson fields: their characteristics, statistics, and simulation.

3.3.2.2 Inhomogeneous Poisson fields

If the point density varies in the area, but the independence property holds analogously to homogeneous Poisson fields, we speak about *inhomogeneous Poisson fields*. Instead of the intensity λ, an intensity function $\lambda(x)$ or an intensity measure Λ is used. Thus, the second property of inhomogeneous Poisson fields should be modified and now corresponds to:

2′. The number of points $N(S)$ in a set S follows the Poisson distribution with parameter $\Lambda(S)$. $\Lambda(S)$ describes the mean of the number of points in the set S. This intensity measure Λ is diffuse; that is, there are no multiple points. If the density $\lambda(x)$ exists with

$$\Lambda(S) = \int_S \lambda(x)dx,$$

then this density is called the intensity function.

Statistics for inhomogeneous fields is more complicated than for homogeneous fields. One problem is the determination of the corresponding intensity function. There are parametric and nonparametric approaches. Parametric approaches are based on the maximum likelihood method, and nonparametric methods use the following simple estimator of the intensity function:

$$\hat{\lambda}_h(x) = \frac{N(b(x, h))}{\pi h^2} \qquad (3\text{-}112)$$

With $b(x, h)$ in (3-112) (b comes from "ball") a disc with the center in x and radius h is denoted. Choosing parameter h usefully is a nontrivial problem. For large h the local fluctuations can be damped and vanish; for small h there are too many local effects. Thus, this is one of those pseudomathematical problems that most likely belong in the category of "philosophical" questions. Without any additional assumptions about the field being considered no final answer can be given. But we can recommend testing different values of h until some plausible results are obtained.

A more general approach uses different kernel functions for estimation of the density function. Finally, we speak about *Matern cluster fields*. Further models of point processes and details about corresponding statistical methods can be found in Stoyan and Stoyan (1994).

3.3.2.3 Matern Cluster Fields

Matern cluster fields are popular and used frequently. The basis of the cluster model is a homogeneous Poisson field of intensity λ_0. The random points of the Poisson field are called parent points. In the second step, daughter points are randomly scattered around each parent point. The union of only daughter points neglecting parent points forms the Neyman-Scott field. We speak about a Matern cluster field if the random number of daughter points in a cluster follows a Poisson distribution with constant parameter μ. In this case the probability function describing the distance of a daughter point scattered in a disc with radius R from the center of a cluster corresponds to

$$P\left(dist \leq r\right) = \begin{cases} \dfrac{r^2}{R^2}, x \leq R \\ 1, otherwise \end{cases} \tag{3-113}$$

The simulation of Matern cluster fields is simple and implemented in most software tools for point processes. For these fields the important characteristics discussed above can be calculated. For example, the K-function yields

$$K\left(r\right) = \pi r^2 + \frac{1}{\lambda_0} \begin{cases} 2 + \frac{1}{\pi}\left[\begin{matrix} \left(8z^2 - 4\right)\arccos z - 2\arcsin z \\ +4z\sqrt{\left(1 - z^2\right)^3} - 6z\sqrt{1 - z^2} \end{matrix} \right], r \leq 2R \\ 1, r > 2R \end{cases} \tag{3-114}$$

$$z = \frac{r}{2R}$$

The pair correlation function [cf. (3-96)] is

$$\rho_1^{(2)}\left(r\right) = 1 + \frac{f\left(r\right)}{2\pi\lambda_0 r}, \quad r \geq 0$$

$$f\left(r\right) = \begin{cases} \dfrac{4r}{\pi R^2}\left[\arccos\left(\dfrac{r}{2R}\right) - \dfrac{r}{2R}\sqrt{1 - \dfrac{r^2}{4R^2}}\right], & r \leq 2R \\ 0, & otherwise \end{cases} \tag{3-114'}$$

Remark: If someone wants to test whether a point pattern should be interpreted as realization of a Matern cluster field or of another well-known model, the application of Monte-Carlo L-tests can be recommended. Simulation algorithms for many popular random point fields are still implemented in conventional statistical software tools such as "R."

Chapter 4
Practical Examples for Mathematical Modeling

4.1 Generalization of Valley Elevation Cross-Profiles

Distinguishing among valleys, or more exactly among types of valleys, is of interest in geomorphology. There are valleys whose primary force of formation was the motion of a glacier. Such valleys tend to follow some shape across the valley, and often one speaks about U-shaped valleys. Other valleys, formed by rivers, for example, tend to be V-shaped. Some of the naive curve fittings described in the literature are generally inadequate, so improvements in the methodology for obtaining and analyzing valley elevation cross-profiles are needed. We propose one such improvement in what follows.

We start with a brief description of meaningful mathematical methods. A problem of generalizing the form of a valley in geomorphology should be translated into an approximation problem in the language of mathematics. Thus, a point cloud should be fitted by piecewise continuous functions.

Hirano and Aniya (1988) led the way to a theoretical discussion concerning the ideal shape of a glaciated valley when they wrote about the so-called catenary curve that can be used for modeling a chain that is supported only at its ends. Power law and parabolic regressions were suggested by Dornkamp and King (1971), James (1996), and others. A further generalization of the power law regression model is the generalized power law (GPL) model described in Pattyn and Van Heule (1998), where they proposed the model $y - y_0 = \alpha |x - x_0|^\beta$. All parameters α, β and the so-called datum (x_0, y_0) should be estimated in this model. The information about curvature is included in parameter β. This model is invariant to changes in coordinate locations, but it cannot be applied directly to asymmetric valleys. Working with the width and depth of a valley Li et al. (2001) solved the problem in their own way. More recent papers are still proposing various generalization methods.

Here we present a special sequential method based on the approximation of a valley elevation cross-profile by using three piecewise continuous functions. The first and the third function come from two different power law regression models. Their parameters β and the mean inclinations of these functions help to describe forms of valley elevation cross-profiles. The second function—if it exists—describes the valley region with minimal inclination, the so-called valley bed, which is modeled by a horizontal line (see Fig. 4.1). Using the second function in this model guarantees an improved accuracy of approximation. But how can we obtain these functions? We apply conventional least-squares techniques for curve fitting and give further details of our special sequential regression method based on fixed and variable point pairs.

More about regression approaches from a mathematical point of view can be found in Draper and Smith (1998). The method of least squares developed by Gauss is widely applied in the geosciences [see Niemeier (2002), Wolf (1979), Reissmann (1976), and the relevant section of Chap. 3]. We do not describe the least-square method in detail here, but rather just start with an introduction to our special technique for cross-profile approximation.

The original data consist of points with coordinates (x, y). The first coordinates x are "relative." They start with value zero and go at right angles to the cross-profile axis. There are identical distances between the neighboring x-coordinates. The second coordinates y are the corresponding valley elevations.

In the first step, the original data set should be divided into three subsets (marked by upper indices). A fixed limit value $\delta > 0$ reflects the allowed depth of the valley bed and should be defined a priori. With the help of this value such points are

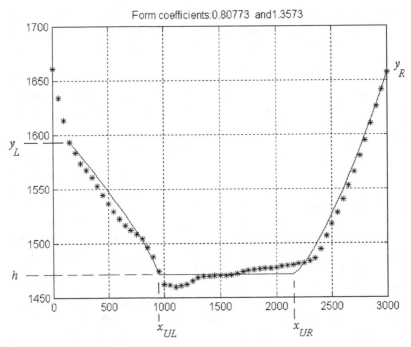

Fig. 4.1 A typical valley elevation cross-profile (*stars*) and its approximation with three continuous functions (**left** and **right** borders are modeled by power law regression functions; the valley bed is given by a linear function)

selected, which are placed in this valley bed. The x-coordinates of separated points remain unchanged. The y-coordinates should be replaced by their mean h.

The second subset has to be continuous, which means that there are no outlier points within a valley bed. The last x-coordinate of the first subset is the neighbor of the first x-coordinate of the second subset and the last x-coordinate of the second subset is the neighbor of the first x-coordinate of the third subset. Initially, the first point of I^1 (and the last point of I^3) is the point at which the local maximum of the corresponding valley elevation nearest to the valley bed is reached. Thus, we have

$$\begin{bmatrix} x_1 \ \dots \ x_N \\ y_1 \ \dots \ y_N \end{bmatrix} \Rightarrow$$

$$\begin{bmatrix} x_1^1 \ \dots \ x_{N_1}^1 \\ y_1^1 \ \dots \ y_{N_1}^1 \end{bmatrix} \cup \begin{bmatrix} x_1^2 \ \dots \ x_{N_2}^2 \\ y_1^2 = h \ \dots \ y_{N_2}^2 = h \end{bmatrix} \cup \begin{bmatrix} x_1^3 \ \dots \ x_{N_3}^3 \\ y_1^3 \ \dots \ y_{N_3}^3 \end{bmatrix} = I^1 \cup I^2 \cup I^3 \ ,$$

$$\begin{bmatrix} x \\ y \end{bmatrix} \in I^1, \quad if \quad x < x_1^2, \qquad \begin{bmatrix} x \\ y \end{bmatrix} \in I^3, \quad if \quad x > x_{N_2}^2 \ , \tag{4-1}$$

$$\begin{bmatrix} x \\ y \end{bmatrix} \in I^2, \quad if \quad |y - \min(y)| \le \delta, \quad y_1^2 = \dots = y_{N_2}^2 = \frac{\displaystyle\sum_{\substack{i=1,\dots N \\ \|y_i - \min(y)\| \le \delta}} y_i}{N_2} = h$$

In the second step, we rely on the following functional structure describing our approximation approach:

$$f_1(x) = \{y_L - h\} \cdot \left(\frac{x_{UL} - x}{x_{UL} - x_L} \right)^{n_L} + h \ ,$$

$$f_2(x) = h,$$

$$f_3(x) = \{y_R - h\} \cdot \left(\frac{x - x_{UR}}{x_R - x_{UR}} \right)^{n_R} + h \ , \tag{4-2}$$

$$x_{UL} = x_1^2, \ x_{UR} = x_{N_2}^2$$

We explain the other parameters from (4-2) below (see also Fig. 4.1). We call the power parameters n_L and n_R form coefficients, and these should be determined using least-square techniques. We make the following model assumptions:

1. A complementary parameter ε should be determined a priori, in order to control the regression accuracy and so dictate the number of loops in our sequential regression approach.
2. There are two kinds of fixed points (x_L, y_L), (x_{UL}, h), and (x_{UR}, h), (x_R, y_R) in the sequential regressions. The points (x_{UL}, h) and (x_{UR}, h) are constant in all loops; the points (x_L, y_L) and (x_R, y_R) can vary.
3. On the "sequential character" of our approach: if the upper limit ε is exceeded, a regression procedure should be repeated with newly calculated input data. In this case the point subsets I^1 (or I^3) should be reduced by omitting the first (or the

last) point, and a new iteration starts. If fewer than five points in a subset remain, then this process should be stopped. The final subsets are denoted by I_1 and I_3. The subset I_1 starts with point (x_L, y_L) and the subset I_3 is finished with point (x_R, y_R) (see Fig. 4.1).

4. At each iteration step, we should solve the following problem in order to obtain the form coefficients:

$$n_L: \quad \sum_{\left[\begin{smallmatrix} x_i \\ y_i \end{smallmatrix}\right] \in I_1} \left(f_1\left(x_i^1\right) - y_i^1 \right)^2 \to \min_{n_L} \,,$$

$$n_R: \quad \sum_{\left[\begin{smallmatrix} x_i \\ y_i \end{smallmatrix}\right] \in I_3} \left(f_3\left(x_i^3\right) - y_i^3 \right)^2 \to \min_{n_R} \tag{4-3}$$

Problem (4-3) can be solved using least-squares techniques, and the solution corresponds to

$$n_L = \frac{\displaystyle\sum_{\left[\begin{smallmatrix} x_i \\ y_i \end{smallmatrix}\right] \in I_1} \ln\left(\frac{x_{UL} - x_i}{p_2}\right) \cdot \ln\left(\frac{y_i - h}{p_1}\right)}{\displaystyle\sum_{\left[\begin{smallmatrix} x_i \\ y_i \end{smallmatrix}\right] \in I_1} \left(\ln\left(\frac{x_{UL} - x_i}{p_2}\right) \right)^2}, \tag{4-4}$$

$$p_1 = y_L - h > 0, p_2 = x_{UL} - x_L > 0$$

and

$$n_R = \frac{\displaystyle\sum_{\left[\begin{smallmatrix} x_i \\ y_i \end{smallmatrix}\right] \in I_3} \ln\left(\frac{x_i - x_{UR}}{p_2}\right) \cdot \ln\left(\frac{y_i - h}{p_1}\right)}{\displaystyle\sum_{\left[\begin{smallmatrix} x_i \\ y_i \end{smallmatrix}\right] \in I_3} \left(\ln\left(\frac{x_i - x_{UR}}{p_2}\right) \right)^2}, \tag{4-5}$$

$$p_1 = y_R - h > 0, p_2 = x_R - x_{UR} > 0$$

The model or generalization accuracy σ can be calculated by

$$\sigma^2 = \frac{1}{N} \left[\sum_{\left[\begin{smallmatrix} x_i \\ y_i \end{smallmatrix}\right] \in I_1} \left(f_1\left(x_i^1\right) - y_i^1 \right)^2 + \sum_{\left[\begin{smallmatrix} x_i \\ y_i \end{smallmatrix}\right] \in I_2} \left(h - y_i^2 \right)^2 + \sum_{\left[\begin{smallmatrix} x_i \\ y_i \end{smallmatrix}\right] \in I_3} \left(f_3\left(x_i^3\right) - y_i^3 \right)^2 \right] \tag{4-6}$$

Example 4.1.1 The following valley elevation cross-profile should be generalized (see Fig. 4.1):

x [m]	0	50	100	150	200	250	300	350	400
y [m]	1660.9	1633.9	1613.7	1593.6	1583.6	1573.8	1567.6	1561.3	1553.1

450	500	550	600	650	700	750	800	850	900
1544.3	1536.6	1529.4	1522.6	1516.5	1512.1	1508.7	1504.1	1496.6	1487.9

950	1000	1050	1100	1150	1200	1250	1300	1350	1400
1474.6	1462.4	1461.4	1460.0	1461.1	1462.4	1465.4	1468.5	1469.2	1469.7

1450	1500	1550	1600	1650	1700	1750	1800	1850	1900
1470.0	1469.9	1470.1	1470.9	1472.2	1474.0	1475.2	1475.6	1476.1	1476.6

1950	2000	2050	2100	2150	2200	2250	2300	2350	2400
1477.1	1477.7	1478.5	1479.4	1480.1	1480.9	1481.8	1483.7	1485.8	1494.7

2450	2500	2550	2600	2650	2700	2750	2800	2850	2900
1507.0	1517.9	1528.0	1540.1	1553.2	1566.1	1580.6	1594.6	1610.3	1626.3

2950	3000
1641.7	1657.0

A complementary parameter ε is determined as follows:

$$\varepsilon = 0.1 \left(y_{\max} - y_{\min}\right) \qquad (*.1)$$

It corresponds to 10% of the absolute elevation difference. In our case $\varepsilon = 20.09$. This parameter helps to control the regression accuracy and thus dictates the number of loops in our sequential regression approach. We set the valley bed height $\delta = \varepsilon$ in (4-1) and obtain the value h.

There are two loops for the regression at the left side and a single loop for the regression at the right side; see model assumptions (1)–(4) described above. We get the following solution of problem (4-3):

$$f_1(x) = \{y_L - h\} \cdot \left(\frac{x_{UL} - x}{x_{UL} - x_L}\right)^{n_L} + h,$$

$$f_2(x) = h, \qquad (*.2)$$

$$f_3(x) = \{y_R - h\} \cdot \left(\frac{x - x_{UR}}{x_R - x_{UR}}\right)^{n_R} + h,$$

with $x_L = 150$, $x_R = 3000$, $x_{UL} = 950$, $x_{UR} = 2150$, $h = 1471.1$,

$y_L = 1593.6$, $y_R = 1657.0$,

$n_L = 0.81$, $n_R = 1.36$

The model accuracy σ given by (4-6) in this case equals (in meters):

$$\sigma = 6.88 \qquad (*.3)$$

Figure 4.2 shows the residuals:

$$res\,(x_i) = \sum_{\left[\begin{smallmatrix} x_i \\ y_i \end{smallmatrix}\right] \in I_1} (f_1\,(x_i) - y_i) + \sum_{\left[\begin{smallmatrix} x_i \\ y_i \end{smallmatrix}\right] \in I_2} (h - y_i) + \sum_{\left[\begin{smallmatrix} x_i \\ y_i \end{smallmatrix}\right] \in I_3} (f_3\,(x_i) - y_i) \qquad (*.4)$$

These residuals can help to analyze the possible locations of the so-called valley ter-
races, which can owe their formation to various factors such as rivers, glaciers, and
so on. Mathematically speaking, these are locations where the residuals change their
signs. The considered cross-profile is nearly V-formed at the left side and nearly
U-formed at the right side (see the form coefficients in Fig. 4.1).

The form parameters, the length of the valley bed, and other parameters can be
taken into account for a further valley cross-profile classification.

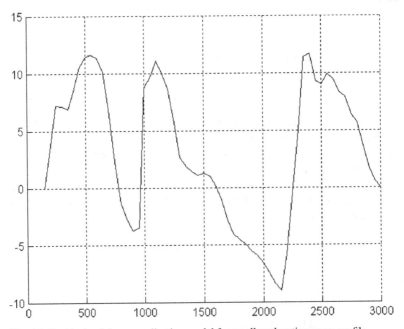

Fig. 4.2 Residuals of the generalization model for a valley elevation cross-profile

4.2 On Fuzzy Propagation of Measurements to Derived Differential-Geometric Characteristics

Real measurements can be considered to be fuzzy values from a mathematical point of view. The uncertainty is the result of incomplete knowledge of "absolute truth." Only value ranges can be proposed in this case. Real measurements provide "raw material" for further research, as they can be interpolated over a grid and approximated by an analytical surface. Moreover, it can be necessary to obtain some surface characteristics of higher order based on the grid. It is clear that any characteristic that is derived from uncertain measurements is also uncertain. In this section we discuss the mathematical modeling of the fuzzy propagation of measurements to derived differential-geometric characteristics.

From a mathematical point of view, most measurements can be assumed to be fuzzy values because absolute precision of a measurement cannot be guaranteed in the real world. There are some statistical tests that are sensitive to the appearance of rough abnormalities in spatial time series [see Waelder (2005a,b)]. Another point of view makes use of the fact that interpolation data are not sets of real numbers but are ranges of values. The distribution within the range may not necessarily be probabilistic. The difference between error and uncertainty is explained in Lodwick and Santos (2003): "Error assumes that a true value exists. Uncertainty denotes incomplete knowledge that is characterized by whether or not one can say that a proposition is exclusively true or false. A statement is uncertain when its (exclusive) truth or falseness can be ascertained."

Uncertainty can be modeled using some useful approaches that have been developed in fuzzy set theory. In *interval arithmetic*, which is one of these approaches, a measurement is considered to be a mathematical object, an interval with two fixed borders: (real) lower- and upper-limit values. The assumed measurement uncertainty can be modeled using the variable width of this interval. It should be noted that an interval is the simplest fuzzy object. A helpful introduction to fuzzy theory can be found in Bandemer and Gottwald (1993).

Usually, measurements should be sampled and transferred into GIS, and generally they are interpolated over a grid. Alternatively, they might be approximated with an analytical surface or some differential-geometric characteristics might be obtained based on sampled measurements. Partial derivatives of higher orders, curvatures, surface maximal shear strain, and surface dilatation are among these characteristics. In all these cases the fuzzy propagation of "original measurements" to derived differential-geometric characteristics is of interest.

Some investigations related to fuzzy surfaces are carried out in Kaleva (1994) and Lodwick and Santos (2003). Here, we discuss a method of the fuzzy propagation of measurements of derived differential-geometric characteristics.

A fuzzy value describes an inaccurately determined or an imprecisely measured value. Fuzzy values can be modeled by convex fuzzy sets. Their characterizing function should have exactly one local maximum [see Bandemer and Gottwald (1993)]. The support of fuzzy intervals corresponds to an interval with real fixed borders

Fig. 4.3 A schematic presentation of two fuzzy intervals $A = \left[a^-, a^0, a^+\right]$ and $B = \left[b^-, b^0, b^+\right]$ with their characterizing functions

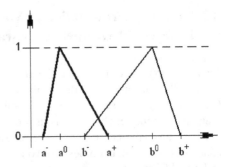

(see Fig. 4.3). Fuzzy interval calculation uses and generalizes methods of interval arithmetic that are utilized in "classical" adjustment theory. Corresponding to the point of view appearing in fuzzy theory, one uses intervals to deal with variable widths that reflect a measure of the uncertainty rather than variances of random values in probabilistic approaches. Further details concerning interval arithmetic can be found in Alefeld und Herzberger (1974). We repeat that in the approach presented in this section, the borders of fuzzy intervals are assumed to be fixed and precise.

In order to explain the relations obtained below, we have to define some basic operations with fuzzy intervals. According to Lodwick and Santos (2003) we ask for only three parameters connected with a fuzzy interval and do not ask for its characterizing function. These three parameters are the lower and upper borders of a fuzzy interval and its so-called *center*: the point within the supporting real interval where the characterizing function has its maximum (see Fig. 4.3).

We denote two fuzzy intervals with $A = \left[a^-, a^0, a^+\right]$ and $B = \left[b^-, b^0, b^+\right]$. The following basic operations with these intervals are necessary: the fuzzy sum $A + B$, the difference $A - B$, the fuzzy product $A \cdot B$, and the fuzzy ratio A/B. A new fuzzy interval $C = \left[c^-, c^0, c^+\right]$ is the result of these operations. For the given parameters, this results in:

$$
\begin{aligned}
C = A + B : \quad & c^- = a^- + b^-, \ c^0 = a^0 + b^0, \ c^+ = a^+ + b^+; \\
C = A - B : \quad & c^- = a^- - b^+, \ c^0 = a^0 - b^0, \ c^+ = a^+ - b^-; \\
C = A \cdot B : \quad & c^- = min\left\{a^- b^-, a^+ b^-, a^- b^+, a^+ b^+\right\}, \\
& c^0 = a^0 b^0, \\
& c^+ = max\left\{a^- b^-, a^+ b^-, a^- b^+, a^+ b^+\right\}; \quad\quad (4\text{-}7) \\
C = A/B : \quad & c^- = min\left\{a^- /b^-, a^+ /b^-, a^- /b^+, a^+ /b^+\right\}, \\
& c^0 = a^0 /b^0, \\
& c^+ = max\left\{a^- /b^-, a^+ /b^-, a^- /b^+, a^+ /b^+\right\} \\
only\ for \quad & b^- > 0 \quad or \quad b^+ < 0
\end{aligned}
$$

Furthermore, we can define the fuzzy square of a fuzzy interval by

$$C = A^2 = A \cdot A:$$

$$\begin{cases} c^- = \min\{a^-a^-, a^+a^+\}, \ c^0 = (a^0)^2, c^+ = \max\{a^-a^-, a^+a^+\}, \\ \qquad a^- \geq 0 \quad or \quad a^+ \leq 0 \\ c^- = 0, \ c^0 = (a^0)^2, c^+ = \max\{a^-a^-, a^+a^+\}, \quad a^- < 0 \ and \ a^+ > 0 \end{cases} \tag{4-8}$$

and the following holds for the fuzzy root:

$$C = \sqrt{A}:$$

$$c^- = \sqrt{a^-}, \ c^0 = \sqrt{a^0}, c^+ = \sqrt{a^+}, \quad for \quad a^- \geq 0 \quad only \tag{4-9}$$

We assume that fuzzy measurements $\tilde{z}_i = [z_i^-, z_i^0, z_i^+]$, $i = 1,\ldots,n$, are given at precise, that is, nonfuzzy points (x_i, y_i), $i = 1\ldots n$. If these measurements have to be interpolated to a nonfuzzy grid $\{X_j, Y_k\}$, $j = 1\ldots N$, $k = 1\ldots M$, then the interpolated elevations $\tilde{Z}_{jk} = [Z_{jk}^-, Z_{jk}^0, Z_{jk}^+]$ also become fuzzy values. Their parameters can be determined depending on the applied interpolation method.

By using distance-dependent weightings of the original data we get

$$\tilde{Z}_{jk} = \alpha_1^{jk} \tilde{z}_1 + \alpha_2^{jk} \tilde{z}_2 + \ldots + \alpha_n^{jk} \tilde{z}_n,$$

$$\sum_{i=1}^{n} \alpha_i^{jk} = 1, \quad \alpha_i^{jk} = Fct\left(d_i^{ij}\right), \quad d_i^{jk} = (x_i - X_j)^2 + (y_i - Y_k)^2, \tag{4-10}$$

$$i = 1\ldots n, \ j = 1\ldots N, \ k = 1\ldots M$$

where the abbreviation "*Fct*" stands for "a function of." If the method of the squares of inverse distances (IDW) is taken into account, (4-10) can be rewritten as

$$\tilde{Z}_{jk} = \alpha_1^{jk} \tilde{z}_1 + \alpha_2^{jk} \tilde{z}_2 + \ldots + \alpha_n^{jk} \tilde{z}_n,$$

$$\alpha_i^{jk} = \frac{w_i^{jk}}{\sum\limits_{i=1}^{n} \left(w_i^{jk}\right)}, \quad w_i^{jk} = \frac{1}{d_i^{jk} + \varepsilon}, \tag{4-11}$$

$$\varepsilon > 0, \ i = 1\ldots n, \ j = 1\ldots N, \ k = 1\ldots M$$

The complementary constant value ε in (4-11) is used in order to avoid uncertainty of weights in the case in which a prediction point coincides with a measurement point. It is often recommended to set $\varepsilon = 0.6$. As the weights in (4-11) are positive, we get the following parameters for interpolated fuzzy heights:

$$Z_{jk}^- = \alpha_1^{jk} z_1^- + \alpha_2^{jk} z_2^- + \ldots + \alpha_n^{jk} z_n^- = \sum_{i=1}^{n} \alpha_i^{jk} z_i^-,$$

$$Z_{jk}^0 = \alpha_1^{jk} z_1^0 + \alpha_2^{jk} z_2^0 + \ldots + \alpha_n^{jk} z_n^0 = \sum_{i=1}^{n} \alpha_i^{jk} z_i^0, \tag{4-12}$$

$$Z_{jk}^+ = \alpha_1^{jk} z_1^+ + \alpha_2^{jk} z_2^+ + \ldots + \alpha_n^{jk} z_n^+ = \sum_{i=1}^{n} \alpha_i^{jk} z_i^+$$

$$i = 1\ldots n, \ j = 1\ldots N, \ k = 1\ldots M$$

Remark 4.2-1 We use small letters to denote grid coordinates $\{X_j, Y_k\}$, $j = 1\ldots N$, $k = 1\ldots M$, in order to avoid unnecessary complexity in the equations that follow.

Remark 4.2-2 If the coordinates of measurements are also fuzzy, one should replace the usual arithmetic operations in (4-11) by the rules in (4-7) and (4-8) for the corresponding fuzzy operations.

In the next step, we can approximate the interpolated fuzzy heights with an analytical function and in this way construct a fuzzy surface. A two-dimensional Lagrange polynom can be used for this approximation, which is constructed using the following (nonfuzzy) Lagrange functions (see Sect. 3.1.1):

$$\varphi_j(x) = \prod_{\substack{i=1\ldots N \\ i \neq j}} \frac{(x - X_i)}{(X_j - X_i)}, \quad \varphi_k(y) = \prod_{\substack{i=1\ldots M \\ i \neq k}} \frac{(y - Y_i)}{(Y_k - Y_i)} \qquad (4\text{-}13)$$

These functions have a well-known property at the grid nodes $\{X_j, Y_k\}$, $j = 1\ldots N$, $k = 1\ldots M$, namely,

$$\varphi_j(x_p) = \delta_{jp} = \begin{cases} 1, \ p = j \\ 0, \ p \neq j \end{cases}$$

$$\varphi_k(y_p) = \delta_{kp} = \begin{cases} 1, \ p = k \\ 0, \ p \neq k \end{cases} \qquad (4\text{-}14)$$

Using the functions in (4-13) we can express the analytical fuzzy surface by

$$\tilde{p}(x,y) = \sum_{\substack{j=1\ldots N \\ k=1\ldots M}} \tilde{Z}_{jk} \varphi_j(x) \varphi_k(y) = \sum_{\substack{j=1\ldots N \\ k=1\ldots M}} \tilde{Z}_{jk} L_{jk}(x,y),$$

$$L_{jk}(x,y) = \varphi_j(x) \varphi_k(y) \qquad (4\text{-}15)$$

We rely on similar deliberations and explanations related to the sign change of $L_{jk}(x,y)$, $j = 1\ldots N, k = 1\ldots M$, as proposed by Lodwick and Santos (2003) in order to obtain the parameters of the fuzzy surface from (4-15) and get:

$$p^-(x,y) = \sum_{\substack{j=1\ldots N \\ k=1\ldots M \\ L_{jk}(x,y) \geq 0}} Z_{jk}^- L_{jk}(x,y) + \sum_{\substack{j=1\ldots N \\ k=1\ldots M \\ L_{jk}(x,y) < 0}} Z_{jk}^+ L_{jk}(x,y),$$

$$p^0(x,y) = \sum_{\substack{j=1...N \\ k=1...M}} Z^0_{jk} L_{jk}(x,y),$$

$$p^+(x,y) = \sum_{\substack{j=1...N \\ k=1...M \\ L_{jk}(x,y) \geq 0}} Z^+_{jk} L_{jk}(x,y) + \sum_{\substack{j=1...N \\ k=1...M \\ L_{jk}(x,y) < 0}} Z^-_{jk} L_{jk}(x,y) \qquad (4\text{-}16)$$

These three functions from (4-16) describe the lower, middle, and upper border surfaces of the fuzzy surface from (4-15).

This basic idea of fuzzy modeling can be used for other differential-geometric characteristics. For example, the gradient length $\tilde{Q}$ of the fuzzy surface in (4-15) can be modeled by

$$\tilde{Q} = \sqrt{\tilde{\Delta}_x^2 + \tilde{\Delta}_y^2} \qquad (4\text{-}17)$$

with

$$\tilde{\Delta}_x = \frac{d}{dx}\tilde{p}(x,y) = \frac{d}{dx}\sum_{\substack{j=1...N \\ k=1...M}} \tilde{Z}_{jk}\varphi_j(x)\varphi_k(y) = \sum_{\substack{j=1...N \\ k=1...M}} \tilde{Z}_{jk}\varphi_k(y)\frac{d}{dx}\varphi_j(x),$$

$$\tilde{\Delta}_y = \frac{d}{dy}\tilde{p}(x,y) = \frac{d}{dy}\sum_{\substack{j=1...N \\ k=1...M}} \tilde{Z}_{jk}\varphi_j(x)\varphi_k(y) = \sum_{\substack{j=1...N \\ k=1...M}} \tilde{Z}_{jk}\varphi_j(x)\frac{d}{dy}\varphi_k(y)$$

$$(4\text{-}18)$$

where it is clear that rules (4-7)–(4-9) should be used.

GIS tools often use finite elements instead of "classical" derivatives. Most filters are constructed using this principle, and after application to a grid, a new—usually correspondingly reduced—grid is obtained. The fuzzy propagation for these procedures can be achieved as follows:

1. Calculation of the Mean

In this method one calculates the following weighted sum (not only with positive weights!):

$$\tilde{m} = \sum_{j,k=1}^{L} \beta_{jk}\tilde{Z}_{jk}, \quad \sum_{j,k=1}^{L} \beta_{jk} = 1 \qquad (4\text{-}19)$$

For the width of the so-called mean window the values $L = 3, 5, \ldots$ are generally used. The parameters for this fuzzy mean (4-19) are

$$m^- = \sum_{\substack{j,k=1 \\ \beta_{jk} \geq 0}}^{L} \beta_{jk} z_i^- + \sum_{\substack{j,k=1 \\ \beta_{jk} < 0}}^{L} \beta_{jk} z_i^+ ,$$

$$m^0 = \sum_{j,k=1}^{L} \beta_{jk} z_i^0 , \tag{4-19'}$$

$$m^+ = \sum_{\substack{j,k=1 \\ \beta_{jk} \geq 0}}^{L} \beta_{jk} z_i^+ + \sum_{\substack{j,k=1 \\ \beta_{jk} < 0}}^{L} \beta_{jk} z_i^-$$

2. Partial Derivatives (Using Finite Elements)

$$\tilde{\Delta}_x^{jk} = \frac{1}{x_{j+1} - x_{j-1}} \left[\tilde{Z}_{(j+1)k} - \tilde{Z}_{(j-1)k} \right], \quad \tilde{\Delta}_y^{jk} = \frac{1}{y_{k+1} - y_{k-1}} \left[\tilde{Z}_{j(k+1)} - \tilde{Z}_{j(k-1)} \right],$$

$$j = 2 \ldots N - 1, \, k = 2 \ldots M - 1 \tag{4-20}$$

The parameters of the fuzzy derivatives with respect to x are

$$\Delta_x^{jk-} = \frac{1}{x_{j+1} - x_{j-1}} \left[Z_{(j-1)k}^- - Z_{(j-1)k}^+ \right], \quad \Delta_x^{jk0} = \frac{1}{x_{j+1} - x_{j-1}} \left[Z_{(j-1)k}^0 - Z_{(j-1)k}^0 \right],$$

$$\Delta_x^{jk+} = \frac{1}{x_{j+1} - x_{j-1}} \left[Z_{(j-1)k}^+ - Z_{(j-1)k}^- \right], \tag{4-20'}$$

$$j = 2 \ldots N - 1, \, k = 2 \ldots M - 1$$

The parameters of the partial derivatives with respect to y can be obtained in a similar fashion.

For the calculation of derivatives of higher order p we use the following recursive rule based on a primary grid of derivatives of order $p - 1$:

$$\tilde{\Delta}_{x(p)}^{jk} = \frac{1}{x_{j+1} - x_{j-1}} \left[\tilde{\Delta}_{x(p-1)}^{(j+1)k} - \tilde{\Delta}_{x(p-1)}^{(j-1)k} \right], \quad \tilde{\Delta}_{y(p)}^{jk} = \frac{1}{y_{k+1} - y_{k-1}} \left[\tilde{\Delta}_{y(p-1)}^{(j+1)k} - \tilde{\Delta}_{y(p-1)}^{(j-1)k} \right] \tag{4-21}$$

It is clear that grids that are built in this way are reduced step by step. The following relations are true for the parameters of (4-21) (derivatives with respect to y can be treated analogously):

$$\Delta^{jk-}_{x(p)} = \frac{1}{x_{j+1} - x_{j-1}} \left[\Delta^{(j+1)k-}_{x(p-1)} - \Delta^{(j-1)k+}_{x(p-1)} \right],$$

$$\Delta^{jk0}_{x(p)} = \frac{1}{x_{j+1} - x_{j-1}} \left[\Delta^{(j+1)0}_{x(p-1)} - \Delta^{(j-1)0}_{x(p-1)} \right], \tag{4-21'}$$

$$\Delta^{jk+}_{x(p)} = \frac{1}{x_{j+1} - x_{j-1}} \left[\Delta^{(j+1)k+}_{x(p-1)} - \Delta^{(j-1)k-}_{x(p-1)} \right]$$

If measurements of two temporally separated epochs (indicated below by [1] and [2]) are compared, then the approaches of the *surface deformation analysis* described in Voosoghi (2000) can be used. We discuss fuzzy propagation for the following characteristics, which are denoted there as *surface dilatation* and (squared!) *surface maximal shear strains*. This "square" can be negative because surface maximal shear strain can take complex values. The general definitions that are given in Voosoghi (2000) can be simplified in our case to:

$$D\tilde{I}L = \tilde{\lambda}_1 + \tilde{\lambda}_2 \text{ and } \tilde{\Gamma}^2 = \left(\tilde{\lambda}_1 - \tilde{\lambda}_2 \right)^2 \text{ with}$$

$$\tilde{\lambda}_1 = 0.5 \cdot \left(PA\tilde{R}_1 + \sqrt{PA\tilde{R}_1^2 - 4 \cdot PA\tilde{R}_2} \right), \tilde{\lambda}_2 = 0.5 \cdot \left(PA\tilde{R}_1 - \sqrt{PA\tilde{R}_1^2 - 4 \cdot PA\tilde{R}_2} \right) \text{ and}$$

$$PA\tilde{R}_1 = tr\left(\tilde{E} \cdot \tilde{A}^{-1} \right), \ PA\tilde{R}_2 = \det\left(\tilde{E} \cdot \tilde{A}^{-1} \right), \text{ where} \tag{4-22}$$

$$\tilde{A} = \begin{pmatrix} 1 + \left(\tilde{\Delta}^{jk}_x [1] \right)^2 & \tilde{\Delta}^{jk}_x [1] \cdot \tilde{\Delta}^{jk}_y [1] \\ \tilde{\Delta}^{jk}_x [1] \cdot \tilde{\Delta}^{jk}_y [1] & 1 + \left(\tilde{\Delta}^{jk}_y [1] \right)^2 \end{pmatrix}, \ \tilde{C} = \begin{pmatrix} 1 + \left(\tilde{\Delta}^{jk}_x [2] \right)^2 & \tilde{\Delta}^{jk}_x [2] \cdot \tilde{\Delta}^{jk}_y [2] \\ \tilde{\Delta}^{jk}_x [2] \cdot \tilde{\Delta}^{jk}_y [2] & 1 + \left(\tilde{\Delta}^{jk}_y [2] \right)^2 \end{pmatrix},$$

$$\tilde{E} = 0.5(\tilde{C} - \tilde{A})$$

It follows from (4-22) that

$$D\tilde{I}L = PA\tilde{R}_1, \quad \tilde{\Gamma}^2 = PA\tilde{R}_1^2 - 4 \cdot PA\tilde{R}_2 \tag{4-22'}$$

After some simplification in (4-22'), one can obtain:

$$D\tilde{I}L = PA\tilde{R}_1 = 0.5 \cdot tr\left(\tilde{C} \cdot \tilde{A}^{-1} - E \right) = 0.5 \cdot tr\left(\tilde{C} \cdot \tilde{A}^{-1} \right) - 1$$

$$= \frac{1}{2 \left(1 + \left(\tilde{\Delta}^{jk}_x [1] \right)^2 + \left(\tilde{\Delta}^{jk}_y [1] \right)^2 \right)}$$

$$\begin{bmatrix} \left(1 + \left(\tilde{\Delta}^{jk}_x [2] \right)^2 \right) \cdot \left(1 + \left(\tilde{\Delta}^{jk}_y [1] \right)^2 \right) - 2\tilde{\Delta}^{jk}_x [2] \tilde{\Delta}^{jk}_y [2] \tilde{\Delta}^{jk}_x [1] \tilde{\Delta}^{jk}_y [1] \\ + \left(1 + \left(\tilde{\Delta}^{jk}_y [2] \right)^2 \right) \cdot \left(1 + \left(\tilde{\Delta}^{jk}_x [1] \right)^2 \right) \end{bmatrix} - 1$$

and

$$PA\tilde{R}_2 = 0.5^2 \cdot \det\left(\tilde{C}\cdot\tilde{A}^{-1} - E\right) = 0.25 \cdot \left(\det\left(\tilde{C}\cdot\tilde{A}^{-1}\right) - tr\left(\tilde{C}\cdot\tilde{A}^{-1}\right) + 1\right)$$

$$= 0.25 \cdot \left(\frac{\det\left(\tilde{C}\right)}{\det\left(\tilde{A}\right)} - tr\left(\tilde{C}\cdot\tilde{A}^{-1}\right) + 1\right) = 0.25 \cdot \frac{\det\left(\tilde{C}\right)}{\det\left(\tilde{A}\right)} - 0.5 \cdot PA\tilde{R}_1 - 0.25$$

$$= \frac{1 + \left(\tilde{\Delta}_x^{jk}[2]\right)^2 + \left(\tilde{\Delta}_y^{jk}[2]\right)^2}{4\left(1 + \left(\tilde{\Delta}_x^{jk}[1]\right)^2 + \left(\tilde{\Delta}_y^{jk}[1]\right)^2\right)} - 0.5 \cdot PA\tilde{R}_1 - 0.25 \tag{4-23}$$

In (4-22′) and (4-23) we apply the rules defined in (4-7)–(4-9).

Now, let us present the fuzzy propagation of measurements of some derived differential-geometric parameters using an example.

Example 4.2.1 We consider the following elevation measurements of a rock glacier [from Waelder et al. (2004)]. The first digits "52" of the y-coordinates are always omitted; all coordinates are rounded):

x [m]	53203	53195	53179	53157	53131	53101	53090	53116	53143
y [m]	12474	12485	12498	12508	12516	12514	12486	12479	12463
z_0 [m]	2382	2382	2381	2383	2385	2386	2389	2390	2390

53174	53192	53177	53144	53123	53089	53061	53045	53069	53106
12455	12427	12396	12405	12424	12440	12441	12405	12403	12386
2390	2390	2400	2401	2400	2399	2398	2405	2406	2408

53136	53162	53172	53166	53130	53087	53056	53040	53055	53067
12373	12364	12344	12288	12313	12333	12349	12313	12307	12296
2406	2410	2415	2427	2424	2425	2417	2434	2437	2440

53078	53116	53051	53036	53016	53007	52995	53011	52824	53377
12259	12215	12109	12121	12135	12148	12192	12167	12046	12701
2449	2457	2511	2510	2510	2511	2501	2499	2619	2332

53241	53304	53198	53377
12439	12466	12639	12701
2385	2392	2341	2332

The elevation measurements $\tilde{z}_i$, $i = 1\ldots n = 43$ should be modeled as fuzzy values with parameters $\tilde{z}_i = \left[z_i^- = z_i^0 - 0.5, z_i^0, z_i^+ = z_i^0 + 0.5\right]$. Here we use the value 0.5 meters as an upper limit for the measurement uncertainty. This value corresponds to the precision that reflects the difficulties connected with data sampling in high mountain regions. The (x,y)-coordinates (Gauss-Krueger coordinate system) are assumed to be fixed and nonfuzzy. Once again let us note that the first digits "52" of the y-coordinate are omitted and all of the coordinates are rounded.

In the first step these measurements should be interpolated with IDW over a grid. Figure 4.4 shows the parameter of the corresponding fuzzy grid values $\tilde{Z}_{jk}$, $i, j = 1 \ldots N = 20$ from (4-11). This parameter is calculated using (4.2). We set $\varepsilon = 0.6$ for the weight calculation. Figures 4.5 and 4.6 show the results of the fuzzy propagation for gradient length using (4-17) and (4-20).

We assume that the (x, y)-coordinates are also fuzzy. Their uncertainty can be described analogously to the elevation measurements, which means that the (x, y)-coordinates can be modeled by $\tilde{u}_i$, $i = 1 \ldots n = 43$, with $\tilde{u}_i = \left[u_i^- = u_i^0 - 0.5, u_i^0, u_i^+ = u_i^0 + 0.5 \right]$, $u = x, y$. Figures 4.7 and 4.8 illustrate some results of the fuzzy propagation.

Comparing Figs. 4.4 and 4.7 and 4.6 and 4.8 reflects the increasing uncertainty of the derived parameters, which is caused by the increased number of fuzzy components that are taken into account in the calculation of the corresponding differential-geometric characteristics.

The application of fuzzy theory is an alternative, meaningful supplement to statistical methods of quality control as well as to the law of error (or variance) propagation from adjustment theory. Generally geoscientific applications use not only a calculated value of a characteristic, but are also concerned with a "confidence interval," which is joined with this characteristic. In particular, the width of this confidence interval plays an important role in further interpretations of obtained

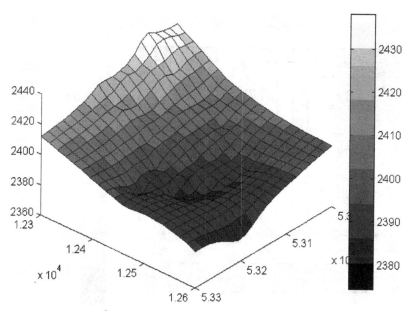

Fig. 4.4 Parameter Z_{jk}^0 of fuzzy values $\tilde{Z}_{jk}$ obtained via (4-11) and (4-12) for the given data. The width of the corresponding fuzzy intervals (in further figures only denoted as "uncertainty width") is constant and equal to 1.0 meter

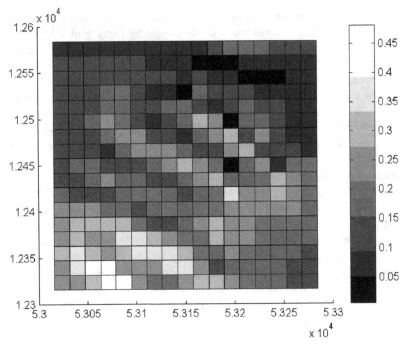

Fig. 4.5 Gradient length

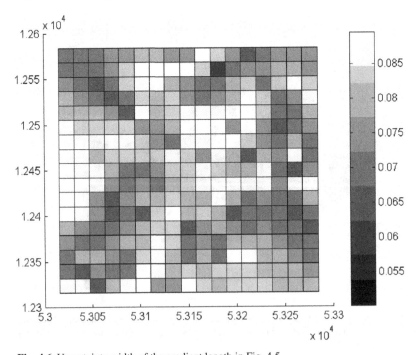

Fig. 4.6 Uncertainty width of the gradient length in Fig. 4.5

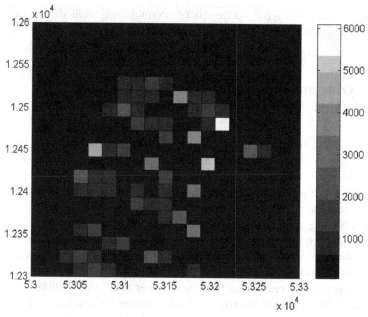

Fig. 4.7 Uncertainty width of $\tilde{Z}_{jk}$ (cf. Fig. 4.4)

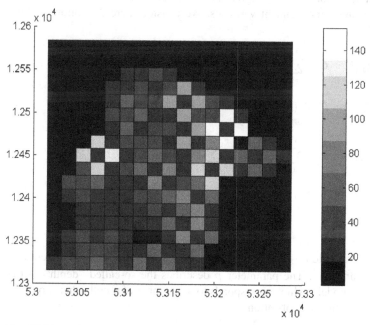

Fig. 4.8 Uncertainty width of the gradient length in Fig. 4.5

results because in some sense it reflects the risk of interpretation. Methods of fuzzy theory can be taken into account for uncertainty modeling and propagation.

4.3 On Analyzing and Forecasting Dam Deformations: Some AR-Models and Their Applications

The movement and deformation of a surface can be considered as a random and dynamic time-varying process. Methods of multiple regression analysis can help to recognize the linear structure of such processes and make the forecast. Qualified monitoring and forecasting of surface movement and deformation can sometimes prevent natural disasters.

A two-step modeling technique is proposed in relation to this problem. Here, we want to discuss some alternative models that can be applied for analyzing the trend of dam deformation as well as for short-term forecasting. These models are based mainly on methods of adjustment theory [see for example Wolf (1979)]. Some ideas come from applied regression analysis [see Draper and Smith (1998)]. The subsection 3.2.1 of this book can be also taken into account. A case study of deformation analysis and short-term forecasting is related to fictive, simulated dam measurements.

We assume two independent variables X, Y (forecasting factors), which describe pressure and temperature and influence dam deformation Z. The restriction of using only two variables is for practical reasons, and, obviously, our models can be simply generalized for more independent variables. We consider here the following four models:

$$Z_1(k) = C + \sum_{i=1}^{p} \alpha_i X(k-i) + \sum_{i=1}^{p} \beta_i Y(k-i) + \varepsilon$$

$$Z_2(k) = C \cdot Z_2(k-1) + \sum_{i=1}^{p} \alpha_i X(k-i) + \sum_{i=1}^{p} \beta_i Y(k-i) + \varepsilon$$

$$Z_3(k) = C + \sum_{i=0}^{p} \alpha_i X(k-i) + \sum_{i=0}^{p} \beta_i Y(k-i) + \varepsilon$$

$$Z_4(k) = C \cdot Z_4(k-1) + \sum_{i=0}^{p} \alpha_i X(k-i) + \sum_{i=0}^{p} \beta_i Y(k-i) + \varepsilon$$

(4-24)

The random variable $\varepsilon \sim N(0, \sigma^2)$ is normally distributed with mean zero and (unknown) variance σ^2. The parameter p describes the so-called "depth" of the recursion of AR-models, which corresponds to the number of months that should be taken into account in our application.

Applying methods from adjustment theory, we can estimate the unknown co-efficients C and α_i, β_i, $i = 0$ $(or\ 1)..p$ using a sufficient number M of equations corresponding to the given measurements $x(j), y(j), z(j)$, $|j| = M$, where M is the number of months over which the measurements are sampled. This number M should satisfy $M \leq 2p + 1$ for models 1 and 2 and $M \leq 2p + 3$ for models 3 and 4. For example, from (4-24) we have for model 1:

$$z_1(p+1) = C + \sum_{i=1}^{p} \alpha_i x(p+1-i) + \sum_{i=1}^{p} \beta_i y(p+1-i) + \varepsilon$$

$$z_1(p+2) = C + \sum_{i=1}^{p} \alpha_i x(p+2-i) + \sum_{i=1}^{p} \beta_i y(p+2-i) + \varepsilon \qquad (4\text{-}25)$$

$$\ldots$$

$$z_1(p+M) = C + \sum_{i=1}^{p} \alpha_i x(p+M-i) + \sum_{i=1}^{p} \beta_i y(p+M-i) + \varepsilon$$

Describing these equations in matrix form as given in (4-25'),

$$\bar{z} = A \cdot \bar{u} + \bar{\varepsilon},$$
$$\bar{z}^T = [z(p+1), \ldots, z(p+M)], \quad \bar{u}^T = [C, \alpha_1, \ldots, \alpha_p, \beta_1, \ldots, \beta_p],$$
$$A(k,1) = 1, \ k = 1..M \qquad (4\text{-}25')$$
$$A(k, l+1) = x(p+k-l), \ k = 1..M, \ l = 1 \ldots p$$
$$A(k, l+p+1) = y(p+k-l), \ k = 1..M, \ l = 1..p$$

we get the well-known solution of (4-25'), which corresponds to

$$\bar{u} = \left(A^T A\right)^{-1} A^T \bar{z} \qquad (4\text{-}26)$$

The accuracy of model fitting can be obtained from

$$\hat{\sigma} = \sqrt{\frac{(A\bar{u} - \bar{z})^T (A\bar{u} - \bar{z})}{M - 2p - 1}} \qquad (4\text{-}27)$$

The statistical goodness of fit can be proved by the empirical mean and correlation coefficient between the real measurements and their forecasted values, which can be calculated using the proposed mathematical models. It should be expected that

$$E(\varepsilon) \sim \hat{\varepsilon} = \bar{z} - A\bar{u} \approx 0$$
$$\hat{\rho} = \rho(A\bar{u}, \bar{z}) \approx 1 \qquad (4\text{-}28)$$

Models 2, 3, and 4 from (4-24) can be handled in an analogous fashion. The value $\hat{\sigma}^2$ from (4-27) can be used as an estimator for the unknown variance σ^2 of $\varepsilon \sim N(0, \sigma^2)$ and applied for the forecasting discussed below.

Example 4.3.1 Let us consider the following simulated dam measurements over 3 years (36 months). Here, pressure and temperature are independent variables X and Y, whereas deformation is a dependent variable Z.

Months 1–12

Def, mm	28.5	60.5	45.0	41.0	52.0	62.0	51.5	38.5	50.0	41.5	70.5	63.0
Pres, mPa	24.2	21.5	23.3	21.2	18.9	21.0	22.3	23.1	20.7	21.9	22.0	22.9
Temp, C	1.5	0.50	12.0	9.0	8.5	3.2	7.1	4.8	10.2	0.2	9.9	1.2

Months 13–24

66.0	54.5	56.5	35.0	85.0	40.0	51.5	55.0	70.5	24.0	33.0	32.5
22.0	21.9	18.8	24.3	20.2	23.0	23.4	21.0	21.9	23.0	22.9	22.9
0.3	4.9	10.1	0.2	0.3	17.9	0.2	0.1	0.3	13.9	20.0	7.8

Months 25–36

26.5	34.5	61.0	55.5	31.5	39.0	53.5	51.0	60.0	24.2	32.9	33.0
22.8	22.3	21.9	21.4	23.3	21.9	22.6	22.0	21.2	24.0	23.0	22.3
8.9	11.9	3.9	18.5	4.2	1.1	0.2	18.1	15.0	14.0	19.0	8.0

These monthly dam deformations are shown in Fig. 4.9.

First, we discuss modeling and forecasting of the deformation for the last four months [$K = 33, 34, 35,$ and 36 from the table above based on measurements obtained during months 1–32; see Fig. 4.9 (left)].

We apply models 1–4 from (4-24) for different values of parameter p and obtain the coefficients of the corresponding multiple linear regression given in Tables 4.2 to 4.5. Figure 4.10 shows a comparison between real measurements and their estimated values.

The forecasting is modeled as follows. First, we use a multiple regression model for the trend prediction. Second, we simulate a normally distributed parameter ε with mean zero and variance $\hat{\sigma}^2$ 100 times and use the obtained maximal *absolute* values for calculating the limits of the forecasting interval. Of course, one can also use the $1.5 \cdot \sigma$ rule for these limits.

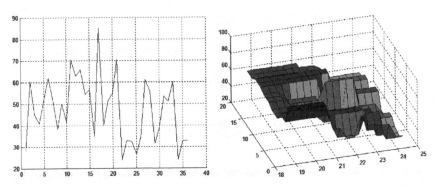

Fig. 4.9 The monthly deformations (**left**) and deformations related to pressure and temperature (**right**)

Table 4.1 Real Measurements and Their Forecasting for the Last Four Months

Month number, K	33	34	35	36
Deformation, *real*, mm	60.0	24.2	32.9	33.0

Table 4.2 The Coefficients of the Multiple Linear Regression and Some Goodness-of-Fit Characteristics [Model 1 from (4-24), Based on Measurements from the First 32 Months]

$p = 2$	$p = 3$	$p = 4$	$p = 5$	$p = 6$	$p = 7$
$C = 96.87$	$C = 55.83$	$C = 57.70$	$C = -76.45$	$C = 73.60$	$C = 147.78$
$\alpha_1 = 0.63$	$\alpha_1 = 0.88$	$\alpha_1 = 1.04$	$\alpha_1 = 2.14$	$\alpha_1 = 2.35$	$\alpha_1 = 4.42$
$\beta_1 = -0.85$	$\beta_1 = -1.78$	$\beta_1 = -0.87$	$\beta_1 = -0.85$	$\beta_1 = -0.74$	$\beta_1 = -0.66$
	$\alpha_2 = -1.17$	$\alpha_2 = -1.81$	$\alpha_2 = -1.68$	$\alpha_2 = -2.20$	$\alpha_2 = -0.91$
$\alpha_2 = -2.21$	$\beta_2 = -1.49$	$\beta_2 = -1.14$	$\beta_2 = -1.44$	$\beta_2 = -1.38$	$\beta_2 = -1.26$
$\beta_2 = -1.10$					
	$\alpha_3 = 1.28$	$\alpha_3 = 1.46$	$\alpha_3 = 2.38$	$\alpha_3 = 1.33$	$\alpha_3 = 1.0$
	$\beta_3 = -0.21$	$\beta_3 = -0.23$	$\beta_3 = -0.4$	$\beta_3 = -0.18$	$\beta_3 = -0.32$
		$\alpha_4 = -0.41$	$\alpha_4 = 0.04$	$\alpha_4 = -0.78$	$\alpha_4 = -1.54$
		$\beta_4 = 0.02$	$\beta_4 = -0.13$	$\beta_4 = -0.06$	$\beta_4 = 0.1$
			$\alpha_5 = 3.79$	$\alpha_5 = 2.39$	$\alpha_5 = 1.35$
			$\beta_5 = -0.45$	$\beta_5 = 0.15$	$\beta_5 = -0.14$
				$\alpha_6 = -3.55$	$\alpha_6 = -5.24$
				$\beta_6 = 0.31$	$\beta_6 = 0.63$
					$\alpha_7 = -3.28$
					$\beta_7 = 0.5$
$\hat{\sigma} = 12.24$	12.83	13.71	12.85	12.56	11.63
$\hat{\varepsilon} = 0.02$ mm $\times 10^{-8}$	0.009	0.004	0.05	0.17	0.07
$\hat{\rho} = 0.62$	0.63	0.63	0.75	0.8	0.87

Table 4.3 The First Coefficient of the Multiple Linear Regression and Some Goodness-of-Fit Characteristics [Model 2 from (4-24), Based on Measurements from the First 32 Months]

$p = 2$	$p = 3$	$p = 4$	$p = 5$	$p = 6$	$p = 7$
$C = 0.1$	$C = 0.01$	$C = -0.006$	$C = 0.01$	$C = 0.42$	$C = 0.30$
$\hat{\sigma} = 12.73$	12.96	13.80	12.97	11.60	11.31
$\hat{\varepsilon} = -2.99$ mm	-1.25	-0.78	0.69	-0.27	-0.33
$\hat{\rho} = 0.57$	0.62	0.63	0.74	0.83	0.88

Table 4.4 The First Coefficient of the Multiple Linear Regression and Some Goodness-of-Fit Characteristics [Model 3 from (4-24), Based Only on Measurements from the First 32 Months

$p = 1$	$p = 2$	$p = 3$	$p = 4$	$p = 5$
$C = 160.0$	$C = 211.6$	$C = 220.4$	$C = 216.5$	$C = 106.3$
$\hat{\sigma} = 10.9$	10.9	11.4	11.9	11.6
$\hat{\varepsilon} = 0.17\,\mathrm{mm} \times 10^{-9}$	-0.04	0.26	-0.33	-0.74
$\hat{\rho} = 0.71$	0.74	0.75	0.77	0.83

Table 4.5 The First Coefficient of the Multiple Linear Regression and Some Goodness-of-Fit Characteristics [Model 4 from (4-24), Based on Measurements from the First 32 Months]

$p = 1$	$p = 2$	$p = 3$	$p = 4$	$p = 5$
$C = 0.25$	$C = 0.17$	$C = 0.20$	$C = 0.22$	$C = 0.21$
$\hat{\sigma} = 12.1$	12.7	12.6	12.6	11.4
$\hat{\varepsilon} = -4.8$	-4.5	-2.7	-1.7	-0.4
$\hat{\rho} = 0.62$	0.63	0.69	0.74	0.83

In Table 4.6 the forecasted values and the sum of the corresponding absolute differences between real measurements (see Table 4.1) and the forecasted values are presented.

From Tables 4.1 and 4.6 it can be seen that long-term forecasting is mostly useless, a fact that is well known in approximation theory. It is more meaningful to

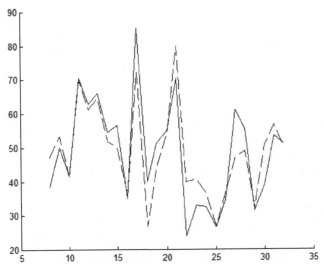

Fig. 4.10 The real monthly deformations (*solid line*) and their forecasting values (*dotted line*) by model 1 from (4-24) with $p = 7$

Table 4.6 Forecasted Intervals and the Sum of the Absolute Differences D between Real Measurements and Centers of Forecasting Intervals Produced by Model 1 from (4-24)

	$K = 33$	$K = 34$	$K = 35$	$K = 36$	d
$p = 2$	45.18 ± 18.4	29.02 ± 18.4	36.80 ± 18.4	26.83 ± 18.4	29.70
$p = 3$	46.34 ± 19.2	29.61 ± 19.2	33.52 ± 19.2	23.99 ± 19.2	28.71
$p = 4$	46.23 ± 20.6	30.44 ± 20.6	33.97 ± 20.6	24.75 ± 20.6	29.34
$p = 5$	41.92 ± 19.3	34.19 ± 19.3	34.54 ± 19.3	24.92 ± 19.3	37.79
$p = 6$	43.88 ± 18.8	36.19 ± 18.8	31.02 ± 18.8	24.80 ± 18.8	38.19
$p = 7$	45.28 ± 17.4	38.22 ± 17.4	36.83 ± 17.4	23.84 ± 17.4	41.84

forecast for only the single month 33 based on the preceding 32 months than for four future months 32–36 at the same time.

Moreover, if additional information about pressure, temperature, and deformation in month 33 is given, one should fit a new multiple regression model based on the months 1–33, and then forecast the value of deformation in month 34 and so on. The choice of a multiple regression model 1–4 from (4-24) should be strongly dependent on expert knowledge concerning the "true nature" of the dam deformation process.

It is clear that other regression models are possible. For example, one can use a generalized model such as (4-28):

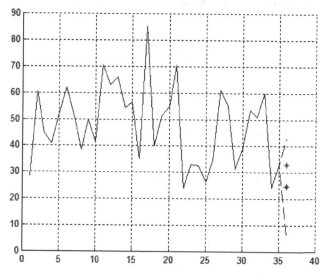

Fig. 4.11 Graphical presentation of the interval-related, short-term forecasting (see the ends of the *dotted lines*) by model 1 from (4-24) with $p = 7$ for the dam deformation in month 36 based on measurements from months 1–35. In this case $\hat{\rho} = 0.84$, $\hat{e} = -0.3 \times 10^{-8}$ and $\hat{\sigma} = 12.2$ are obtained. The real deformation in month 36 is drawn as the *upper star*. The center of its forecasting interval is drawn as the *lower star*

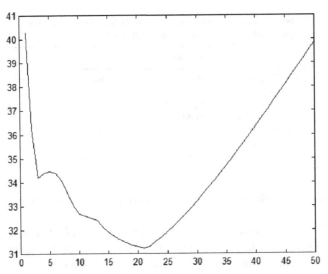

Fig. 4.12 Dependence of values d on the choice of *eps* for the multiple regression model in (4-29)

$$Z_5(k) = C + \sum_{i=0}^{p} \alpha_i \frac{X(k-i)}{Y(k-i) + eps} + \sum_{i=0}^{p} \beta_i Y(k-i) + \varepsilon \qquad (4\text{-}29)$$

The parameter *eps* helps to correct zeros of temperatures. We use $eps = 21$ here because this value leads to the minimum d (see Fig. 4.12). In this case the following

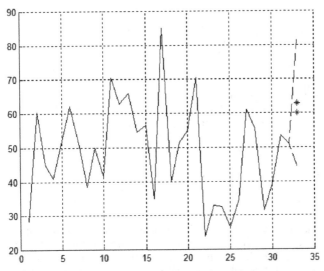

Fig. 4.13 Graphical presentation of the interval-related, short-term forecasting (*dotted lines*) by the model given in (4-24). The real deformation in the month 33 is drawn as the *lower star*. The center of its forecasting interval is drawn as the *upper star*

results are obtained (cf. Table 4.6): $\hat{\rho} = 0.54$, $\hat{\varepsilon} = -0.07\,\text{mm} \times 10^{-10}$, and $d = 31.2$. The parameter $p = 2$ leads to $\hat{\sigma} = 12.5$.

Figure 4.12 shows the values of d depending on the choice of *eps* in (4-29). We can see that the optimal value corresponds to *eps* ≈ 21. Figure 4.13 presents the result of forecasting for month 33 using this model (cf. Fig. 4.11).

Finally, there can never be "point-exact" forecasting because of unknown, future process oscillations caused by the random parameter ε. Graphically, interval-related, short-term forecasting can be done as shown in Figs. 4.11 and 4.13, and some numerical formulas can be found in Chap. 5.

Chapter 5
Some Code Examples

Here we present some (pseudo) code examples that have developed for the methods proposed in this volume. The numbers of the corresponding equations are given in brackets. The various steps are numbered for simplicity. We mark comments within the code with double "!." These code examples can easily be adapted to any programming language.

Some General Designations

ai or a(i)	i^{th} element of vector a
Aij or (i,j)	element of matrix A corresponding to the ith line and j^{th} column
inverse(A)	inverting procedure of a matrix A
square root(a)	square root procedure should be used
Sum(a)	sum of elements of a vector a
SumColumn(A)	a function with an input matrix leading to line vectors containing the sum of the elements in each column of the input matrix A
A1=A(:,k:num_col)	this means a matrix that starts with the kth column of A
a1 =A(:,k)	this means the kth column of A
A2=A(k:num_lin,:)	this means a matrix that starts with the kth line of A
a2=A(k,:)	this means the kth line of A
AT= transpose(A)	procedure transposing a matrix
A*b	product of a matrix A with a vector b
Abs(b)	procedure for calculating the absolute value of b
mean(a)	mean calculation procedure for data set a
var(a)	variance calculation procedure for data set a
A{i,j}	i,j^{th} element of A (A is a matrix of matrices, for example)
SearchMin(a)	procedure leading to the minimal value of matrix/vector a
SearchMax(a)	procedure leading to the maximal value of matrix/vector a

Length(a)	length calculation procedure leading to the number of elements of a vector a
SearchWhereMin(b)	procedure leading to the position of minimal value of a vector b in b
SearchWhereMax(b)	procedure leading to the position of maximal value of a vector b in b

Generalized Arithmetical Mean (3–2)

```
1. Read xi,yi,zi, i=1...N
2. Read point of prediction (xo,yo)
3. Distance vector Di(power of two of distance)
epsilon = 0.6 (!is often recommended!)
For i=1 to N with step = 1
  Di=(xi−x0)²+(yi−y0)²
  wDi=1/(Di+epsilon)
End (i)
4. Sum of distances
S=0
For i=1 to N with step = 1
  S = S+wDi
End (i)
5. Weights Wi for zi, prediction value z0
z0 = 0
For i=1 to N with step=1
  If (Di=0)
    Wi=1
    z0=z
  Stop
  Else
    Wi=wDi/S
    z0=z0+zi*Wi
  End (if)
End (i)
```

Remark: Set $yi = 0$ for the one-dimensional case.

Lagrange Interpolation Method (3–5)

```
1. Read xi,yj,zij, i=1...N, j=1...M (measurements over
a mesh)
2. Calculation of the Lagrange function Fi_xi at any
point xi
```

```
For i=1 to N with step=1
  Fi_xi=1
  For k=1 to N with step=1
    If(k<>i)
      Fi_xi=Fi_xi*(x−xk)/(xi−xk)
    End (if)
  End (k)
End (i)
```

3. Calculation of the Lagrange function Fi_yj at any point yi

```
For j=1 to M with step=1
  Fi_yj=1
  For k=1 to M with step=1
    If(k<>j)
      Fi_yj=Fi_yj*(y−yk)/(yj−yk)
    End (if)
  End (k)
End (j)
```

4. Calculation of the value of the two-dimensional Lagrange polynomial at any point (x,y)

```
Lij=0
For i=1 to N with step=1
  For j=1 to M with step=1
    Lij=Lij+zij*Fi_xi*Fi_yj
  End (j)
End (i)
```

Remark: Set $M = 1$ and all Fi_yj equal to 1 for the one-dimensional case.

1D-Cubic Splines (3–6)

```
1. Read xi,zi, i=1...N
2. Calculating distances h(j)=hj
For j to (N−1) with step=1
 h(j+1)=x(j+1)−x(j)
End (j)
3. Calculating lambda(j), mu(j) and d(j)
For j=2 to (N−1) with step
  lambda(j)=h(j+1)/(h(j)+h(j+1))
  mu(j)=1−lambda(j)
  d(j)=(6/(h(j)+h(j+1)))*((z(j+1)−z(j))/
  h(j+1)−(z(j)−z(j−1))/h(j))
End (j)
```

4. Creation of MAT and of v: the matrix and the vector
  from (3−8)

```
MAT(i,j)=0
For i to (N−2) with step=1
  For j to (N−2) with step 1
    If (i=j)
      MAT(i,j)=2
    End (if)
    If (j=i−1) and (i>1)
      MAT(i,j)=mu(i+1)
    End (if)
    If (j=i+1) and (i<N−2)
      MAT(i,j)=lambda(i+1)
    End (if)
  End (j)
  v(i)=d(i+1)  (!This should be a column−vector!)
End (i)
```

5. Calculation of the second derivatives M(j) from (3−7)

```
M(1)=0
M(N)=0
IMAT=inverse(MAT)
M(2 to N−1)=IMAT*v
```

6. Calculation of the parameters A(j) and B(j)

```
For j=1 to (N−1) with step 1
  A(j)=(z(j+1)−z(j))/h(j+1)−h(j+1)*(M(j+1)−M(j))/6
  B(j)=z(j)−M(j)*(h(j+1))²/6
End (j)
```

7. Spline calculation Sxx at any point xx

```
If (xx >= x(j)) and (xx <= x(j+1))
  Sxx=M(j)*(x(j+1)−xx )³/(6*h(j+1))+
      M(j+1)*(xx−x(j))³/(6*h(j+1))+
      A(j)*(xx−x(j))+B(j)
End (if)
```

2D-Polynomial Regression (3–11)

1. Read xi,yi,zi, i=1...N
2. Set K and L
3. Prove that (K+1)*(L+1)<=N
4. Generation of LSE from (3−11)

```
For kstar=0 to K with step=1
  For lstar=0 to L with step=1
    num_line=(lstar+1)+kstar*(L+1)
```

```
    For k=0 to K with step=1
      For l=0 to L with step=1
        num_column=(l+1)+k*(L+1)
        help_sum1=0
        help_sum2=0
        For i=1 to N with step=1
          xx=xi
          yy=yi
          zz=zi
          help_sum1=help_sum1+xx^(kstar+k)*yy^(lstar+l)
          help_sum2=help_sum2+zz*xx^(kstar)*yy^(lstar)
        End (i)
        MAT(num_line,num_column)=help_sum1
        v(num_line)=help_sum2
      End (l)
    End (k)
  End (lstar)
End (kstar)
5. Solution of LSE from (3-11): the coefficients akl
IMAT=inverse(MAT)
a=IMAT*v
(!v is a column-vector. The result is a column-vector
corresponding to: a00,a01,...,a0L,a10,a11,...,a1L,...,
aK0,aK1,...,aKL!)
```

Remark: Set all $yi=1$ and $L=0$ for the one-dimensional case. For large N, K, and L as well as for "unfortunately chosen" locations of measurements some numerical problems are possible.

B-Curve (3–12)

```
1. Read xi,yi,zi, i=0...N
2. Set a parameter t
3. Calculation of the value Bt of B-curve at t
hbx=xi (!a vector!)
hbx=yi
hbz=zi
For r=1 to N with step=1
  For i=0 to (N-r) with step=1
    bx(i)=(1-t)*hbx(i)+hbx(i+1)*t
    by(i)=(1-t)*hby(i)+hby(i+1)*t
    bz(i)=(1-t)*hbz(i)+hbz(i+1)*t
  End (i)
```

```
  hbx=bx
  hby=by
  hbz=bz
End (r)
Bt=(hbx,hby,hbz)
```

B-Surface (3–13)

```
1. Read grid data xij,yij,zij, i,j=0...N
2. Set two parameters u and v
3. Calculation of the value Buv of B-surface at (u,v)
hbx=xij (!a matrix!)
hbx=yij
hbz=zij
For r=1 to N with step=1
  For i=0 to (N−r)with step=1
    For j=0 to (N−r)with step=1
      bx(i,j)=hbx(i,j)*(1−u)*(1−v)+hbx(i,j+1)*(1−u)*v
             +hbx(i+1,j)*u*(1−v)+hbx(i+1,j+1)*u*v
      by(i,j)=hby(i,j)*(1−u)*(1−v)+hby(i,j+1)*(1−u)*v
             +hby(i+1,j)*u*(1−v)+hby(i+1,j+1)*u*v
      bz(i,j)=hbz(i,j)*(1−u)*(1−v)+hbz(i,j+1)*(1−u)*v
             +hbz(i+1,j)*u*(1−v)+hbz(i+1,j+1)*u*v
    End (j)
  End (i)
hbx=bx
hby=by
hbz=bz
End (r)
Buv=(hbx,hby,hbz)
```

Hair Wavelet Family (3–62)

```
1. Input arguments: a,b,x
2. Output argument: yWav
3. Function ''hair wavelet'':
Function yWav=FunWaveHair(a,b,x)
yWav=0.0
yy=0
t=(x−b)/a
```

```
If (t<0.5) and (t>=0)
  yy=1.0
End
If (t>=0.5) and (t<1.0)
  yy=-1.0
End
a=square root(a)
yWav=yy*1/a
End Function
```

Moving Average (3–75)

```
1. Read ti,zi, i=1...N
2. Set m<(N-1)/2
3. Generation of new time series
For i=(m+1)to(N-m)with step=1
  help_sum=0
  For j=-m to m with step=1
    help_sum=help_sum+z(i+j)
  End (j)
z_aver(i)=help_sum/(2*m+1)
End (i)
```

Influence Function (3–91)

```
1. Read xi,yi,zi(marks), i=1...n
2. Set parameters R and M (Par_M here)
3. Generation of the distance matrix D, matrices A, Sdd;
   matrices of matrices AA, sd
For i=1 to n with step=1
 num_nb = 0;
  For j = 1 to n with step=1
    If (j<>=i) Then Do
      D(i,j)=(x(i)-x(j))²+(y(i)-y(j))²
      di=D(i,j)
        If (di<=R)
        num_nb=num_nb+1
          For kk=1 to M with step=1
            A(num_nb,kk)=zj*di^(kk-1)
            AA{i}=A
```

```
              Sdd(num_nb,kk)=di^(kk-1)
              sd{i}=sdd
           End (kk)
        End (if-di)
      End (if-j)
    End (j)
End (i)
```

4. LSE and its solution: vector a=[1,akoef]

```
For i=1 to n with step=1
  AQ=AA{i}
  SQ=sd{i}
  vz=SumColumn(AQ)
  vz0=z(i)*SumColumn(SQ)
  koef{i}=vz-vz0; (!koef is matrix of matrices!)
End (i)

For i=1 to n with step=1
  ko=koef{i}
  For ij=1 to Par_M with step=1
    For j=1 to Par_M with step=1
    M(i,j)=ko(ij)*ko(j)
    End (j)
  End (ij)
End (i)
```

M2=M(:,2:Par_M) (!This means a matrix that starts with the second column of M!)
vec2=-M(:,1) (!This means the first column of M with changed signs!)

MM=transpose(M2)*M2

vecM=transpose(M2)*vec2
akoef=inverse(MM)*vecM

5. Model accuracy (sigma)

```
kumul=0
For i=1 to n with step=1
  s_gew=0
  s_aver=0
  num_nb=0
  For j=1 to n with step=1
    If (j<>i)
      di=d(i,j)
      If (di<=R)
        help=1
        num_nb=num_nb+1
```

```
        For kk=2 to Par_m with step=1
          help=help+akoef(kk-1)*di^(kk-1)
        End (kk)
        NA(num_nb)=z(j)*help
        s_gew=s_gew+help
        s_aver=s_aver+NA(num_nb)
      End (If-di)
    End (If-j)
  End (j)
  kumul=kumul+(z(i)*s_gew-s_aver)^2
End (i)

kumul=kumul/n

sigma=square root(kumul)
```

Epanecnikov Kernel Function from (3–98)

```
1. Input arguments t,h
2. Output argument yFunEpan
Function yFunEpan=FunEpan(t,h)
z=0
If (t<=h) and (t>=−h)
  z=(3/4)*(1/h)*(1−t^2/h^2)
End (If)
yFunEpan=z
End Function
```

Calculation of the Section of Two-Dimensional Windows from (3–98): Function FunWindowSection

```
1. Input parameters trans,WindowWidth,WindowHeight
2. Output parameters yFunSec
3. Function
yFunSec=FunWindowSection(trans,WindowWidth,
WindowHeight)
z=0
abst1=Abs(trans(1))
abst2=Abs(trans(2))
  If (abst1<=par1) and (abst2<=par2)
    diff1=par1−abst1
```

```
      diff2=par2-abst2
      z=diff1*diff2
   End (If)
yFunSec=z
End Function
```

Estimation of the Mark Correlation Function (3–100′)

```
1. Read data xi,yi,mi(marks), i=1...n
2. Set parameter h, the size of the observation window:
its height is called WindowHeight, its width is denoted
with WindowWidth
3. Which procedure should be chosen for weight
calculation?
Set Kern=2 if (3-98) and Kern=1 if (3-99)
```

```
norm_fact=mean(m2)
norm_fact=norm_fact²
num00=0
```

```
For i=1 to n with step=1
   For j=1 to n with step=1
      d(i,j)=(x(i)-x(j))²+(y(i)-y(j))²
      d(i,j)=square root(d(i,j))
      korwolke(i,j)=m(i)*m(j) (!If the function (a) in
      (3-100)is used. Else use |m(i)-m(j)| for (b) in (3-100)!
      End (j)
End (i)
```

```
dd=2*h
```

```
xmin=SearchMin(d)
xmax=SearchMax(d)
```

```
ymin=SearchMin(korwolke)
ymax=SearchMax(korwolke)
```

```
rInt=DivideInterval([xmin,xmax]) with step=dd
(!rInt is vector [xmin, xmin+dd, xmin+2*dd,...,xmax-dd,
xmax]!)
```

```
num=Length(rInt)
```

```
If (Kern=1)
   For nn=1 to num with step=1
      sum0=0
```

```
    k=0
    naechstr=xmin+(nn-1)*dd
    r(nn)=naechstr
    For i=1 to n with step=1
      For j=1 to n with step=1
        If (d(i,j)>=naechstr-dd/2) and
(d(i,j)<=naechstr+dd/2)
          sum0=sum0+korwolke(i,j)
          k=k+1
        End (If-d)
      End (j)
    End (i)
    If (k=0)
      yy(nn)=99999 (!or another mark for indefinite
      values!)
    Else
      yy(nn)=(1/k)*sum0
    End (If-k)
  End (nn)
End (Kern)

If (Kern=2)
  For nn=1 to num with step=1
    sum0=0
    k=0
    gew=0
    naechstr=xmin+(n-1)*dd
    r(nn)=naechstr
    For i=1 to n with step=1
      For j=1 to n with step=1
        If (d(i,j)>=naechstr-h) and (d(i,j)<=naechstr+h)
          trans=[x(j)-x(i),y(j)-y(i)]
help=FunWindowSection(trans,WindowWidth,WindowHeight)
          (!use function from 5.11!)
          gewhelp=0
          Ab=square root ((x(j)-x(i))²+(y(j)-y(i))²)
          (!use the corresponding square root
          procedure!)
          Ab=Ab-naechstr
            If (help>0.00000000001)
              gewhelp=FunEpan(Ab,Par_Ep)/help
              (!use function for (3-98)!)
            End (If-help)
          sum0=sum0+korwolke(i,j)*gewhelp
          gew=gew+gewhelp
```

```
        k=k+1
      End (If-d)
    End (j)
  End (i)
  If (gew=0)
    yy(nn)=99999 (!or another mark for indefinite
    values!)
  Else
    yy(nn)=sum0/gew
  End (If-gew)
  End (nn)
End (Kern)
```

```
Mark_cor_func=yy/norm_fact
```

Estimation of the L-Function (3–109′)

```
1. Read data xi, yi, i=1...n
2. Calculation of the intensity and the distance matrix
```

```
xmin=SearchMin(x)
xmax=SearchMax(x)
ymin=SearchMin(y)
ymax=SearchMax(y)
```

```
intensity=n/((xmax-xmin)*(ymax-ymin))
```

```
For i=1 to n with step=1
  For j=1 to n with step=1
    d(i,j)=(x(i)-x(j))²+(y(i)-y(j))²
    d(i,j)=square root(d(i,j))
  End (j)
End (i)
```

```
3. Calculation of the minimal and maximal values of the
distance matrix d
```

```
rmin=SearchMin(d)
rmax=SearchMax(d)
```

```
4. The choice of the parameter dd: How fine should the
L-function be drawn?
```

```
r=DivideInterval([rmin,rmax]) with step=dd
(!Here, the corresponding procedure should be used!)
```

5. Calculation of the functions L and K

```
len=Length(r)
For k=1 to len with step=1
  summe=0
  For i=1 to n with step=1
    For j=1 to n with step=1
      If (d(i,j)<=r(k)) and (i is not equal to j)
        summe=summe+1
      End (j)
    End (i)
      K(k)=summe/n
      K(k)=K(k)/intensity
      L(k)=square root(K(k)/pi)
      End (if-d)
End (k)
```

Valley Form Generalization, Equations (4–1)–(4–6)

1. Read data xi, yi, i=1...n, d=x(i+1)-x(i) should be constant for each i=1...n-1

2. A dividing into three groups

```
ybigest=SearchMax(y)
y=y-ybigest
(!In the further procedure y-values are assumed to be
negative. Finally, a re-calculation yreal will be made!)
yreal=y+ybigest
```

3. Dividing into three subsets

```
Depth=10
(!This is a constant value describing the depth of the
valley bed. It can be defined as absolute value like
here or relatively to the depth of the valley, for
example 10% of the valley depth!)

ymin=SearchMin(y)
num_min=SearchWhereMin(y)
xmin=x(num_min)

k1=0
k2=0
k3=0
kk3=0
```

```
For i=1 to n with step=1
  diffy=Abs(y(i)−ymin)
  If (diffy<=Depth)
    kk3=kk3+1
    xx3(kk3)=x(i)
    yy3(kk3)=y(i)
  End (if−diffy)
End (i)

x3min=SearchMin(xx3)
x0=x3min

For i=1 to n with step=1
  diffy=Abs(y(i)−ymin)
  If (diffy<=Depth) and (x(i)−x0<=d)
    k3=k3+1
    x3(k3)=x(i)
    y3(k3)=y(i)
  End (if−diffy)
  If (k3>0)
    x0=x(k3)
  End (if−k3)
End (i)

x3left=x3(1)
x3right=x3(k3)

For i=1 to n with step=1
  If (x(i)<x3left)
    k1=k1+1
    x1(k1)=x(i)
    y1(k1)=y(i)
  Else If (x(i)>x3right)
    k2=k2+1
    x2(k2)=x(i)
    y2(k2)=y(i)
  End (if)
End (i)

len3=Length(y3)
mean_value=Sum(y3)/len3

For i=1 to len3 with step=1
  yd3(i)=mean_value
End (i)

xd3=x3

wert=SearchMax(y1)
```

```
num=SearchWhereMax(y1)
x1=x1(num:k1)
y1=y1(num:k1) (!Overwriting these vectors: Only the
elements from ''num'' up to the end of vectors are
remained!)
k1=Length(x1)
wert=SearchMax(y2)
num=SearchWhereMax(y2)
x2=x2(1:num)
y2=y2(1:num) (!Overwriting these vectors: only the
elements up to ''num'' are remained!)
k2=Length(x2)
```

4. The choice of eps1 and eps2 controlling the accuracy of the valley form generalisation at the left and right side

```
eps1=50
eps2=eps1
```
(!Here, these constants are equal and given as absolute values. One can also use different accuracies or relative values depending on the depth of the valley!)

5. Call the subprocedure **ProcTalKontur**

Subprocedure **ProcTalKontur**

5.1 Setting some additional parameters

```
KritWertLeft=99999
KritWertRight=99999
AnzLoops=20
loops1=0
loops2=0
x1h=x1
x2h=x2
y1h=y1
y2h=y2
lenh1=Length(x1h)
lenh2=Length(x2h)
IndLoopsLeft=1
IndLoopsRight=1
```

5.2 Calculation the power function on the left side

```
If (lenh1>0)
  While (IndLoopsLeft=1)
    loops1=loops1+1
    XLL=x1h(1)
```

```
    XUU=x3(1)
    Y01=-y1h(1)
    HH=-mean_value
(! This value was still calculated in 3.!)
    p1=HH-Y01
    p2=XUU-XLL
    p3=HH
    sum1=0
    sum2=0
    help1=0
    help2=0
    For j=1 to lenh1 with step=1
      If (p2*p1<>0)
        help1=Log((XUU-x1h(j))/p2)
(!Use the corresponding procedure for calculating the
logarithm here!)
          hhh=p3+y1h(j)
          If (Abs(hhh)<0.00001)
            hhh=0.00001
          End (if-Abs)
          help2=Log(hhh/p1)
      End (if-p2*p1)
      sum1=sum1+help1*help2
      sum2=sum2+help1²
    End (j)
    NL=99999
    If (sum2<>0)
      NL=sum1/sum2
    End (if-sum2)
    xd1=DivideInterval([x1h(1),XUU]) with step=d
    len1=Length(xd1)
    help3=99999
    For j=1 to len1 with step=1
      yd1(j)=-HH
      If (XLL<XUU)
        help3=(XUU-xd1(j))/(XUU-XLL)
        If (NL>0)
          help3=help3^NL
        Else
          help3=0
        End (if-NL)
      End (if-XLL)
      yd1(j)=-HH+(HH-Y01)*(help3)
    End (j)
```

5.3 Controlling the accuracy and setting of new fixed points

```
    genleft=yd1(1:Length(y1h))-y1h
(!Use the corresponding procedure for generating the
subvector yd1 (1:Length(y1h))!)
    wertL=SearchMax(Abs(genleft))
    numL=SearchWhereMax(Abs(genleft))
(!Finding the position of the maximal value: x1h
re-starts with this value!)
    KritWertLeft=wertL
    Ind1=(KritWertLeft>eps1) and (lenh1-numL+1>5)
(!This variable ''Ind1'' gets 0 if false and 1 if true!)
    If (Ind1=1)
       x1h=x1h(numL:lenh1)
       y1h=y1h(numL:lenh1)
(!Use the corresponding procedure for the subvector
generation!)
       lenh1=Length(x1h)
    Else
       IndLoopsLeft=0
    End (if-Ind1)
    IndLoopsLeft=(IndLoopsLeft=1) and (loops1<AnzLoops)
(!This variable gets 0 if false and 1 if true!)
   End (While)
End (if-lenh1)
```

5.4 Calculation of the power function on the right side

```
If (lenh2>0)
  While (IndLoopsRight=1)
    loops2=loops2+1
    XRR=x2h(lenh2)
    XUU=x3(k3)
    Y02=-y2h(lenh2)
    HH=-mean_value
    p1=HH-Y02
    p2=XRR-XUU
    p3=HH
    sum1=0
    sum2=0
    help1=0
    help2=0
    For j=1 to lenh2 with step=1
      If (p2*p1<>0)
        help1=Log((x2h(j)-XUU)/p2)
```

```
        hhh=p3+y2h(j)
        If (Abs(hhh<0.00001))
(!Use the corresponding procedures ''Abs'' and ''Log''
meaning ''absolute value'' and ''logarithm'' here!)
          hhh=0.00001
        End (Abs)
        help2=Log(hhh/p1)
      End (If-p2*p1)
      sum1=sum1+help1*help2
      sum2=sum2+help1²
      End (j)
NR=99999
      If (sum2<>0)
        NR=sum1/sum2
      End (sum2)
      xd2=DivideInterval([XUU,x2h(lenh2)]) with step=d
(!Use the corresponding procedure for interval
dividing!)
      len2=Length(xd2)
      help3=99999
      For j=1 to len2 with step=1
        yd2(j)=-HH
        If (XRR>XUU)
          help3=(xd2(j)-XUU)/(XRR-XUU)
          If (NR>0)
            help3=help3^NR
          Else
            help3=0
          End (if-NR)
        End (if-XRR)
        yd2(j)=-HH+(HH-Y02)*(help3)
      End (j)
```

5.5 Control of the accuracy and setting new fixed points

```
      genright=yd2(2:Length(y2h)+1)-y2h
(!Use the corresponding procedure for generating the
subvector yd2(2:Length(y2h))!)
      wertR=SearchMax(Abs(genright))
(!Here, the corresponding procedure should be used!)
      numR=SearchWhereMax(Abs(genright))
(!Finding the position of the maximal value: x2h
re-finishes with this value!)
      KritWertRight=wertR
      Ind2=(KritWertRight>eps2) and (numR>5)
(!This variable gets 0 if false and 1 if true!)
```

```
      If (Ind2=1)
        x2h=x2h(1:numR)
        y2h=y2h(1:numR)
        lenh2=Length(x2h)
      Else
      IndLoopsRight=0
      End (Ind2)
      IndLoopsRight=(IndLoopsRight=1) and
(loops2<AnzLoops)
(!This variable gets 0 if false and 1 if true!)
    End (While)
End (if-lenh2)
```

Some Operations with Fuzzy Values: Chapter 4

Some operations with fuzzy values are described here. The following designations are used in the corresponding functions: Input arguments: two-dimensional vectors par1, par2, par3 (par1 describes the left limits of both fuzzy values, par3 describes the right limits of both fuzzy values, the locations of the maxima of both characterizing functions are given in par2) Output arguments: a three-dimensional vector (interval form of a fuzzy value)

```
1. The fuzzy sum of two fuzzy values

Function yFunSum=FunFuzzySum(par1, par2, par3)
z=[99999,99999,99999]
(!One can also use another ''marker''!)
z(2)=par2(1)+par2(2)
z(1)=par1(1)+par1(2)
z(3)=par3(1)+par3(2)
yFunSum=z
End Function

2. The fuzzy difference of two fuzzy values

Function yFunDif=FunFuzzyDif(par1,par2,par3)
z=[99999,99999,99999]
(!One can also use another ''marker''!)
z(2)=par2(1)-par2(2)
z(1)=par1(1)-par3(2)
z(3)=par3(1)-par1(2)
yFunDif=z
End Function
```

3. The fuzzy product of two fuzzy values

```
Function yFunProd=FunFuzzyProd(par1,par2,par3)
z=[99999,99999,99999]
(!One can also use another ``marker''!)
z(2)=par2(1)*par2(2)
z(1)=SearchMin([par1(1)*par1(2),par1(1)*par3(2),
par3(1)*par1(2),
     par3(1)*par3(2)])
z(3)=SearchMax([par1(1)*par1(2),par1(1)*par3(2),
par3(1)*par1(2),
     par3(1)*par3(2)])
yFunProd=z
End Function
```

4. The fuzzy dividing of two fuzzy values

```
Function yFunSub=FunFuzzySubt(par1,par2,par3)
z=[99999,99999,99999]
(!One can also use another ``marker''!)
If (par1(2)>0)
  z(2)=par2(1)/par2(2)
  z(1)=SearchMin([par1(1)/par1(2),par1(1)/par3(2),
     par3(1)/par1(2),par3(1)/par3(2)])
  z(3)=SearchMax([par1(1)/par1(2),par1(1)/par3(2),
     par3(1)/par1(2),par3(1)/par3(2)])
End (if-par1)
If (par3(2)<0)
  z(2)=par2(1)/par2(2)
  z(1)=SearchMin([par1(1)/par1(2),par1(1)/par3(2),
     par3(1)/par1(2),par3(1)/par3(2)])
  z(3)=SearchMax([par1(1)/par1(2),par1(1)/par3(2),
     par3(1)/par1(2),par3(1)/par3(2)])
End (if-par3)
yFunSub=z
End Function
```

5. The fuzzy power of two (square) for a fuzzy value

```
Function yFunQuad=FunFuzzySquare(par1,par2,par3)
z=[99999,99999,99999]
(!One can also use another ``marker''!)
z(2)=par2(1)*par2(2)
z(1)=SearchMin([par1(1)*par1(2),par1(1)*par3(2),
par3(1)*par1(2),
     par3(1)*par3(2)])
z(3)=SearchMax([par1(1)*par1(2),par1(1)*par3(2),
par3(1)*par1(2),
```

```
        par3(1)*par3(2)])
If (z(1)<0)
  z(1)=0;
End (if-z)
yFunQuad=z
End Function
```

AR-Model (1) from (4–24)

1. Read data set S

2. The numbers of lines and columns of the matrix S are called ``len'' and ``br''. S(:,b) are all elements of this matrix in the column b.

```
Z=S(:,1); (! deformation, mm!)
X=S(:,2); (! pressure, mpa !)
Y=S(:,3); (! temperature, C !)
press_min=SearchMin(X)
press_max=SearchMax(X)
temp_min=SearchMin(Y)
temp_max=SarchMax(Y)
```

3. Modeling

```
len_wahr=len
len=len-4
(!The last four months should be predicted!)
ANTW=[] (!A cleared matrix!)
ANT_V=[]
KOEF=[]
CORMAT=[]
KK=3
m=KK-1 (! The choice of parameter m=p from (4-24). For
this parameter m>(len-3)/3 should hold!)
A=Zeros(len-m,2*m+1)
(! A matrix with (len-m) lines and (2*m+1) columns
filled with zeros !)
v=Zeros(len-m,1)

For i=1 to (len-m) with step=1
  counter_i=i+m
  A(i,1)=1
  v(i)=Z(counter_i)
  For j=2 to (m+1)with step=1
```

```
      counter_j=(j-1)
      A(i,j)=X(counter_i-counter_j)
  End (j)
  For j=(m+2) to (2*m+1) with step=1
      counter_j=(j-(m+3)+1)
      A(i,j)=Y(counter_i-counter_j)
  End (j)
End (i)

unbek=inv(transpose(A)*A)*(transpose(A)*v)
C=unbek(1)
alpha=Zeros(1, m)
beta=Zeros(1, m)

For k=2 to Length(unbek) with step=1
  If (k<=m+1)
  alpha(k-1)=unbek(k)
  Else
    beta(k-(m+1))=unbek(k)
  End (If)
End (k)

KOEF=unbek
genau=transpose(A*unbek-v)*(A*unbek-v)
e_dach=Sum(A*unbek-v)
[lA,bA]=SizeOf(A)
(! Use the corresponding procedure for determining the
size of a matrix here!)
mit_gen(KK)=square root(genau(KK)/(lA-bA))
```

3. Prediction of the deformation for the last four
months via ``VZ'' by the model (1) and its comparison
with the true values called ``wahre''

```
VZ=[]
If (len_wahr>len)
  For num=(len+1) to len_wahr with step=1
    ii=num-len
    VZ(ii)=C
    For k=1 to m with step=1
      VZ(ii)=VZ(ii)+alpha(k)*X(num-k)+beta(k)*Y(num-k)
    End (k)
  End (num)
End (If)

wahre=transpose(Z(from (len+1) to (len_wahr)))
ANT_V=VZ-wahre
abs_sum=Sum(Abs(transpose(ANT_V)))
```

Conclusion

In closing, I want to explain the solution of the puzzle about sixteen corners of a four-dimensional cube mentioned in Chap. 1. The solution can be found in two different ways.

The first way is "mathematical." The four-dimensional space has four basis vectors. Thus, there are four coordinate axes. Let our cube be a unit cube for simplification with one corner at the origin $(0,0,0,0)$ of the coordinate system. The other corners of the cube are $(0,0,0,1)$, $(0,0,1,0)$, $(0,0,1,1)$, ... ,$(1,1,1,1)$, so we have exactly four places for setting 0 or 1. Thus, we have $2^4 = 16$ possibilities. QED

The second way is the "engineer's" solution. We imagine the four-dimensional cube, which is possible if we can understand dimension as a time axis. A four-dimensional unit cube is a three-dimensional cube that was moved during a time unit. Like a photograph shot by a camera with a delayed action we see two three-dimensional cubes at the same time. One three-dimensional cube has eight corners and two cubes have sixteen such corners. QED

Is there a reader who still doubts that four-dimensional cubes exist? I hope not.

Bibliography

Alefeld, G., and Herzberger, J. (1974), *Einführung in die Intervallrechnung*. Bibliograph. Institut, Mannheim

Bandemer, H., and Gottwald. S (1993), *Einführung in die Fuzzy-Methoden*. Akademie Verlag GmbH, Berlin

Box G. E. P., and Jenkins, G. M. (1976), *Time Series Analysis, Forecasting and Control*. Holden-Day, San Francisco

Bracewell, R. (1978), *The Fourier Transform and Its Applications*. 2nd Edition, McGraw-Hill, New York

Brockwell, P. J., and Davis, R. A. (1991), *Time Series: Theory and Methods*. 2nd Edition, Springer, Berlin/Heidelberg/New York

Chiang, C. L. (1980), *An Introduction to Stochastic Processes and Their Applications*. Krieger, Huntington, New York

Chung, K. L. (1968), *A Course in Probability Theory*. Harcourt, Brace, and World, New York

Dierckx, P. (1993), *Curve and Surface Fitting with Splines*. Clarendon Press, Oxford

Doornkamp, J. C., and King, C. A. M. (1971), *Numerical Analysis in Geomorphology*. Edward Arnold, London

Draper, N. R., and Smith, H. (1998), *Applied Regression Analysis*, Wiley-Interscience, New York

Farin, G. (1993), *Curves and Surfaces for CAGD: A practical Guide*. 3rd Edition, Academic Press, New York

Fowkes, N. D., and Mahony, J. J. (1994), *An Introduction to Mathematical Modelling*. John Wiley & Sons, Ltd.

Fowler, A. C. (1997), *Mathematical Models in the Applied Sciences*. Cambridge University Press

Hamming, R. W. (1973), *Numerical Methods for Scientists and Engineers*. 2nd Edition, McGraw-Hill, New York

Hardy, G. H. (1992), *A Mathematician's Apology*. Canto Edition, Cambridge University Press

Hersh, R. (1997), *What Is Mathematics, Really?* Oxford University Press, New York/ Oxford

Hirano, M., and Aniya, M. (1988), A rational explanation of cross-profile morphology for glacial valleys and of glacial valley development. *Earth Surface Processes and Landforms* **13**: 707–716

James, L. A. (1996), Polynomial and power functions for glacial vallye cross-section morphology. *Earth Surface Processes and Landforms* **21**: 413–432.

Kaleva, O. (1994) Interpolation of fuzzy data. *Fuzzy Sets and Systems* **61**: 63–70

Kallenberg, O. (1986), *Random Measures*. Academic Press, London

Karr, A. F. (1986), *Point Processes and Their Statistical Inference*. Marcel Dekker, New York

Li, Y., Liu, G., and Cui, Z. (2001), Longitudinal variations in cross-section morphology along a glacial valley: A case-study from the Tien Shan, China. *Journal of Glaciology* **47** (157): 243–250

Lodwick, W. A., and Santos, J. (2003), Constructing consistent fuzzy surfaces from fuzzy data. *Fuzzy Sets and Systems* **135**: 259–277

Niemeier, W. (2002), *Ausgleichungsrechnung. Eine Einführung für Studierende und Praktiker des vermessungs- und Geoinformationswesens*. Walter de Cruyter, Berlin/New York

Pattyn, F., and Van Heule, W. (1998), Power law or power flaw? *Earth Surface Processes and Landforms* **23**: 761–767

Plutchik, R. (1968), *Foundations of Experimental Research*. Harper & Row, New York/Evanston, IL/London

Reissmann, G. (1976), *Die Ausgleichungsrechnung*. VEB Verlag für Bauwesen, Berlin

Stoyan and Stoyan (1994), *Fractals, Random Shapes and Point Fields*. John Wiley & Sons, Chichester, UK

Strang, G., and Nguyen, T. (1997), *Wavelets and Filter Banks*. Wellesley-Cambridge Press, Wellesley, MA

Voosoghi, B. (2000), Intrinsic deformation analysis of the earth surface based on 3-dimensional displacement fields derived from space geodetic measurements. Dissertation, University of Stuttgart

Wackernagel, H. (1995), *Multivariate Geostatistics*. Springer, Berlin/Heidelberg/New York

Waelder, O. (2005a), Ein empirisches statistisches Verfahren zur Indikation der Datenabnormitäten mittels spezieller Wavelet-Splines. *Austrian Journal of Statistics* **34**(3): 251–261

Waelder, O. (2005b), A method for sequential thinning of digital raster terrain models II: mixed locally adaptive wavelet-splines and anisotropy. *Photogrammetie-Fernerkundung-Geoinformation* **2**: 123–127

Waelder, O., Krainer, K, and Mostler, W. (2004), Praktische Anwendung von speziellen Spline-Verfahren zur Gletscherkinematik am Beispiel eines aktiven Blockgletschers (Tirol). *Österreichische Zeitschrift für Vermessung und Geoinformation* **3/4**: 107–117

Wickerhauser, M. V. (1994), *Adapted Wavelet Analysis from Theory to Software*. A.K. Peters, Ltd., Wellesley, MA

Wolf, H. (1979), *Ausgleichungsrechnung*. Ferd. Dümmlers Verlag, Bonn

Yaglom, A. M. (1986), *Correlation Theory of Stationary and Related Random Functions*. Springer, Berlin/Heidelberg/New York

Index